SpringerBriefs in Biochemistry
and Molecular Biology

For further volumes:
http://www.springer.com/series/10196

Gerhard Bauer • Joseph S. Anderson

Gene Therapy for HIV

From Inception to a Possible Cure

 Springer

Gerhard Bauer
University of California Davis
Sacramento, CA, USA

Joseph S. Anderson
University of California Davis
Sacramento, CA, USA

ISSN 2211-9353   ISSN 2211-9361 (electronic)
ISBN 978-1-4939-0433-4 ISBN 978-1-4939-0434-1 (eBook)
DOI 10.1007/978-1-4939-0434-1
Springer New York Heidelberg Dordrecht London

Library of Congress Control Number: 2014930434

Preface

Combination antiretroviral therapy (ART) for HIV has, without a doubt, saved many lives of people infected with HIV over the last decades, but we are still left with the devastating memory of the large number of deaths caused by HIV and AIDS during the 1980s and early 1990s. To this day, there is still no cure for HIV, and there is also no vaccine to reliably prevent HIV infection. The World Health Organization estimated the number of people infected with HIV worldwide in 2010 at 34 million with about 2.7 million newly infected individuals. This may be an underestimate, however, as many cases go unreported. The total number of AIDS deaths in 2010 was estimated to be around 1.8 million. One of the drawbacks of ART is the fact that it needs to be taken under high compliance lifelong; otherwise the virus will rebound, and drug-resistant mutants can arise. Additionally, many patients suffer from side effects caused by antiretroviral medication, ranging from mild to severe, sometimes limiting quality of life. Another aspect is the high cost of lifelong drug treatment and the limited ability to make these drugs available in developing countries, very much affected by HIV and AIDS. Taken together, these facts have been motivating us and many of our colleagues to continue working towards a cure for HIV that we hope we can elicit with stem cell gene therapy for HIV.

In 1988 an article published by David Baltimore in the journal *Nature* introduced the term "intracellular immunization," which meant engineering HIV target cells to become resistant to HIV by the insertion of "anti-HIV genes." Since these early times, when gene therapy was a very new and emerging research field, members of our research group have been working on the development of such a gene therapy application, particularly stem cell gene therapy for HIV. In spite of drawbacks and funding problems, this research has continued to this day, with current developments promising to come much closer to a cure for HIV than ever before.

Several years ago, the idea of a functional cure for HIV was postulated. This was precipitated by an interesting, anecdotal clinical case in Berlin, Germany. An HIV-positive patient with leukemia had to receive an allogeneic bone marrow transplantation to successfully cure his leukemia. The bone marrow donor had been specifically selected for this patient, not only for the tissue match but also for being homozygous for the CCR5 deletion, a natural chemokine receptor deletion on blood cells, including HIV target cells, that leads to HIV resistance. The phenomenon that a small number of people living in central and northern Europe are carrying this deletion without an adverse phenotype and exhibit natural resistance to HIV is well known. CCR5 acts as the secondary receptor for macrophage tropic strains of HIV, and the absence of this receptor restricts HIV from attaching to and entering HIV target cells. Additionally, most initial HIV infections occur through macrophage tropic strains. After the allogeneic bone marrow transplantation, the patient's leukemia was cured, as expected, but also another remarkable phenomenon occurred: For the last 7 years, there has been no detectable HIV viral load in this patient, under complete ART withdrawal. This case suggests that the transplantation of HIV-resistant hematopoietic stem cells was able to generate an HIV-resistant immune system that has been able to control HIV replication for several years. This transplant is very similar to what has been attempted in stem cell gene therapy clinical applications, and by applying the optimal settings, we believe that the outcome seen in the "Berlin patient" can be repeated using engineered autologous hematopoietic stem cells in other HIV-infected individuals.

In this book, we describe the individual aspects of gene therapy for HIV, from its early development to our current knowledge, from early clinical trials to current and planned future clinical applications, set out to possibly cure HIV. It is the authors' belief that stem cell gene therapy for HIV, if proven successful, can be commercialized and made affordable. Automated, closed system culture systems could be developed, and there is no real technical limit to bringing such systems to developing nations in an easy-to-use application, as long as there is enough effort made to actually develop this. A true impact could also be made if gene therapy vectors could be developed that target hematopoietic stem cells in vivo, making them resistant to HIV. Additionally, with this book, it is our sincere wish to also inspire young researchers to take on the interesting field of gene therapy for HIV and help bring about a long-needed cure for the disease.

This book would not have been possible without decades of hard work from many noted researchers in this field. We therefore would like to particularly acknowledge the contributions of Dr. Donald Kohn, Dr. John Zaia, and Dr. John Rossi. The first potent anti-HIV genes applied clinically and the first stem cell gene therapy clinical trials for HIV were initiated by them at City of Hope and Children's Hospital Los Angeles, among them was the first case of a pediatric clinical trial of stem cell gene therapy for HIV. We also would like to thank our colleague Steve Tobin for his valuable input on this manuscript.

Contents

Chapter 1
Principles of Gene Therapy

Gene therapy, in essence, is defined as the insertion, removal, or manipulation of one or multiple genes inside a cell to treat a specific disease. Most of the time, however, gene therapy simply means "Gene transfer for therapeutic purposes." This can be accomplished by transient or permanent insertion of genes. If permanent insertion is required, this can be accomplished by either targeting a particular locus within the genome or allowing the gene to be transferred and integrated into the genome in a random fashion. In the past, the term "gene therapy" has also been widely applied to basic laboratory research geared towards a better understanding of or the development of more efficient methods to transfer genes. However, such research does not really constitute gene therapy as no therapy for a disease is really being developed. Gene therapy can be applied in animals and in humans. Specific animals with gene defects similar to defects found in humans serve as relevant animals for human gene therapy applications. For the purpose of this chapter, we will limit our discussion to the topic of human gene therapy and will discuss animal models in a later chapter.

Often, gene therapy is applied as a technique for correcting a defective gene responsible for disease development in somatic cells. An early but still quite common form of gene therapy in humans involves applications in monogenic diseases, where disease is caused by the absence of a functional gene product normally produced by a single gene. In such a case, gene therapy can be carried out by the permanent insertion of this particular gene into a nonspecific location within the genome of a suitable target cell to replace the function of the defective gene. While the nonfunctional gene remains in place, the new, "randomly" integrated gene now acts as a "normal" functional gene and results in clinical improvement of the condition. The inserted gene is usually at all times fully upregulated by the use of a strong promoter, and insertional mutagenesis can be a concern due to possible gene integration right next to an oncogene. This will be discussed more in depth in later chapters. Nevertheless, several applications of this method have already produced cures in patients with monogenic diseases without serious adverse events in the majority of the cases [1]. Other forms of gene therapy may involve the

G. Bauer and J.S. Anderson, *Gene Therapy for HIV: From Inception to a Possible Cure*, SpringerBriefs in Biochemistry and Molecular Biology, DOI 10.1007/978-1-4939-0434-1_1, © Gerhard Bauer and Joseph S. Anderson 2014

correction of a particular mutation in a defective gene by use of homologous recombination. This method is much more sophisticated since the manipulated gene remains in the locus within the genome where it is actually supposed to reside, under endogenous control. However, this method is complicated and quite ineffi-cient but promises the huge advantage of normal control of gene expression and restoration of gene function in a more physiological way.

In theory, other than somatic cell gene therapy, germ line gene therapy could also be pursued. Germ cells could theoretically be modified by the insertion and permanent integration of therapeutic genes into their genomes. This would lead to a genetic modi-fication in the germ cells that could be passed onto the offspring and also to later gen-erations. This approach would allow for the permanent correction of an inherited genetic disease, but would not cure an adult patient suffering from this disease. Additionally, this approach could result in unwanted genetic modifications and would also be highly prone to abuse. Obvious ethical reasons, therefore, warranted the making of laws prohibiting germ cell gene therapy in human applications [2].

Somatic cell gene therapy is much less problematic. Therapeutic genes are transferred only into cells of the body, and great care is taken that genes are not transmitted into germ cells. Genetic modifications will be restricted to the targeted somatic cells and to the individual treated with this particular form of gene therapy [2]. Regulatory agencies require strict tests that provide evidence that only somatic cells were targeted by a gene therapy approach and no transmission of genetic modification into the germ line occurred.

Injection of Naked DNA

This is the simplest method used to transfer a gene [3]. For clinical gene therapy applications, naked plasmid DNA generated in bacteria and then highly purified is suspended at a high concentration in a clinical grade buffer called a "vehicle." This suspension is then directly injected into a patient, often intramuscularly. Clinical trials of naked plasmid DNA injection showed only moderate success as gene expression was low and transient. Some clinical trials used naked DNA PCR products, with similar low and transient gene expression [4]. These clinical trials clearly demonstrated that uptake of naked DNA by cells is rather inefficient. However, some DNA vaccines still apply intramuscular injection of naked DNA in conjunction with adjuvants to prime the immune system during short-term gene expression to invoke a vigorous immune response. Recently, some improvement in naked DNA transfer has been achieved by the in vivo application of electroporation [5] or also previously using the "gene gun," which utilizes DNA-covered gold par-ticles that are "shot" into cells using high-pressure gas [6].

Naked DNA transfer belongs into the category of "nonviral gene transfer." As will be discussed later, viral gene transfer methods are currently widely applied to increase the efficiency of gene transfer. However, nonviral gene transfer methods such as naked DNA transfer do offer some advantages. They allow for relatively noncomplicated large-scale production of bacterial plasmid preparations in fermentors followed by affinity column purification and endotoxin removal. If these plasmid preps are used on cells cultured in vitro, then some more efficient transfection methods (as outlined below) can be applied. Recent advances in plasmid vector technology have also led to methods that allow for gene transfer and gene expression efficiencies similar to those achieved by viruses.

Nonviral Gene Transfer (Transfection) Methods

The term "transfection" is used when a nonviral form of gene transfer is applied. Electroporation is a transfection method that uses short pulses of high voltage applied to cultured cells and recently also to cells in vivo. This method quickly and transiently opens pores in the cell membrane so DNA in an appropriate buffer can enter the cytoplasm. If applied properly, this method can be quite efficient and can be applied to a wide range of cells, including quiescent cells. However, low cell viability, dependent on the cell type, is experienced frequently. This is particularly true with isolated primary cells. A modification of the electroporation technique is the "electron-avalanche transfection" method which applies a high-voltage plasma in very short pulses. Good transfection efficiency with much higher cell viability has been achieved using this method [7].

The "Gene Gun": As briefly stated above, this method utilizes DNA-coated gold particles that are "shot" into cells using a pulse of high-pressure gas. The coated gold particles slowly release the DNA into the cytoplasm. A human gene therapy clinical trial was carried out using this device; however, only low transfection and gene expression efficiencies were observed [6].

Sonoporation is a method that uses ultrasound of a certain frequency and amplitude that is directed towards the cultured cells. The energy in the sound waves is thought to break up parts of the cell membrane, causing "acoustic cavitation." During cell membrane disruption, DNA can migrate into the cytosol [8].

Magnetofection is a method which utilizes magnetic nanoparticles that bind DNA [9]. Cells are cultured in a tissue culture dish or flask. A strong magnet is then placed under the culture vessel to draw the DNA-coated magnetic nanoparticles onto the cells and immobilize them there. Nanoparticles can then be passively taken up by the cells with subsequent release of the DNA. Transfection efficiency, however, remains low.

Lipoplexes and Polyplexes: Anionic and neutral lipids were initially applied to produce so-called lipoplexes, micelles capable of carrying plasmid DNA vectors [10]. These lipids are relatively nontoxic and do not harm cultured cells. Therefore, they can potentially be used for large-scale clinical grade transfections of even primary cells. They yield high transfection efficiencies with excellent cell viability; however, they are difficult to manufacture.

Cationic lipids have nowadays almost completely replaced anionic or neutral lipids as their method of production is much less complicated. Negatively charged

DNA can also easily be packaged into positively charged "liposomes," small vesicles formed by these lipids. Liposomes are transported through the cell membrane by endocytosis, allowing the encapsulated DNA to be moved into the cytoplasm efficiently [11]. Unfortunately, in the beginning, transfection efficiency was still low, as the release of the DNA from the liposome was not efficient and subsequent degradation of DNA in the lysosome occurred where these liposomes would be transported to. To improve transfection efficiency, electroneutral lipids were added to form "lipoplexes," and "fusogenic lipids" were developed which allowed DNA to be released from the liposomes prior to degradation in the lysosomes. Cationic lipids have been used in human clinical gene therapy application such as cancer gene therapy clinical trials [12] and also in gene therapy for the treatment of cystic fibrosis where respiratory epithelial cells were the target. What should be pointed out here, however, is the fact that cationic lipids do produce dose-limiting toxicities that need to be carefully monitored to avoid cell death.

Dendrimers are highly branched macromolecules with surfaces that allow for many different properties, depending on the way they are constructed. A positively charged dendrimer, for instance, can be used to bind negatively charged DNA or even RNA increasing its stability during gene transfer. The macromolecule is moved through the cell membrane by endocytosis. Dendrimers can be engineered to a specific size and to target a specific cell, and they may be less toxic than cationic lipids; they have recently been applied in preclinical HIV gene therapy experiments [13]. Manufacturing conditions for dendrimers are challenging and currently not easy to overcome as slow and expensive reaction steps are involved.

Viral Gene Transfer (Transduction) Methods

The term "transduction" is used when a virus-mediated form of gene transfer is applied.

All viruses have in common that they need to bind to their host cell and introduce their genetic material into the cell. The viral genetic material contains information how to produce more viral copies and how to utilize the cell's machinery for this purpose. The host cell will carry out the instructions encoded by the viral genes and produce a multitude of new virus which gets released from the cell, spreading to other cells and infecting them. Some types of viruses insert their genes into the host's genome, while most others do not. There are two main types of viral infections: lytic and lysogenic infections. After introducing the viral genetic information into the target cell, viruses of the lytic type quickly produce more viruses, burst from the cell, and infect more cells. This causes a quick onset of cell destruction and pathology in an organism. Generally, lysogenic viruses integrate their genetic information into the DNA of the host cell and may live dormant in an organism for a prolonged period of time. Integrated viral DNA does get multiplied in the same fashion as the cell's DNA; however, full expression of the viral genes leading to viral replication continues to be suppressed. When triggered, full viral gene expression and replication occurs, with subsequent destruction of the host cell and development of pathology.

Retroviruses

The genetic material in retroviruses is RNA which is surrounded by a capsule of proteins that is designed to protect the viral RNA from degradation and lends stability to the viral particle. Additionally, retroviruses coat themselves with a layer of cell membrane that they carry with them from the cell they replicated in and budded from. Through this layer of cell membrane, the viral envelope protrudes, which allows the particle to bind to the appropriate target cell in a highly specific manner. When a retrovirus infects a target cell, it fuses with the cell membrane, enters the cytoplasm, uncoats, and introduces its RNA together with unique viral enzymes such as reverse transcriptase and integrase, into the target cell. Immediately after entering the cell, reverse transcription occurs, the process that transcribes the viral RNA into double-stranded DNA utilizing the viral reverse transcriptase. In the case of oncoretroviruses, which are very simple retroviruses, the finished DNA has to wait for the breakdown of the nuclear membrane during cell division so it can enter the nucleus. The Moloney leukemia virus (MLV) belongs to this group of retroviruses. Lentiviruses, more complicated retroviruses possessing regulatory and accessory genes, utilize cellular transport mechanisms to transport their newly synthesized DNA into the nucleus through the intact nuclear membrane. Lentiviruses can infect and integrate into nondividing cells. HIV belongs into the group of lentiviruses. DNA in the cell nucleus must then be integrated into the host genome. This task is accomplished by the viral enzyme integrase. Quasi random integration of retroviral DNA is carried out. Both oncoretroviruses and lentiviruses integrate into active genes. Oncoretroviruses, however, prefer start regions of active genes. After the genetic material has been integrated, it can be replicated in its integrated form during cell division without harm to the cell. Integrated viruses can also be triggered to replicate inside the infected cell causing the generation of a multitude of viral particles destroying the host cell in the process. HIV, for instance, almost immediately starts to produce a burst of viruses that infects and destroys other cells until the host organism's immune system starts to suppress viral replication. Integrated viruses, such as oncoretroviruses and lentiviruses, cannot be eliminated from the infected host organism and remain with the organism lifelong.

Both oncoretroviruses (predominantly MLV) and lentiviruses (predominantly HIV) have been engineered to serve as viral gene therapy vectors. Remark: In the literature, oncoretroviral vectors are often referred to as "retroviral" vectors, while HIV-based vectors are referred to as "lentiviral" vectors.

As a gene therapy vector, these viruses have made replication incompetent since they do not carry the viral genome in the viral particle. Instead, the viral genes have been replaced with a gene to be transferred into the host cell. Additionally, vector particles are usually pseudotyped which means that their viral envelope is replaced with another envelope allowing to regulate vector particle cell tropism. For instance, lentiviral vectors are regularly pseudotyped with the vesicular stomatitis virus G glycoprotein (VSV-G) envelope, which allows the vector particle to bind to and infect almost all vertebrate cells. The vector particle carries the gene of interest as two copies of RNA, in conjunction with all viral enzymes needed for all viral functions to be carried out. The vector particle is produced in a producer cell, which is programmed to generate the viral proteins and the gene of interest in the form of a messenger RNA that can be packaged into the vector particle. The original retroviral RNA, however, is not packaged into the new particle since it is lacking the vital packaging sequence.

Both oncoretrovirus- and lentivirus-based vectors have the great advantage of efficient gene transfer and permanent gene integration into the host cell genome with efficient, prolonged gene expression. Gene integration normally does not lead to disruption of cell function, particularly when the integrated gene copy number is limited to one or two. The disadvantages of oncoretroviral vectors are as follows. These vectors can only integrate into dividing cells, and they have a tendency of being silenced in the long run, particularly in vivo. Additionally, they tend to integrate into active start regions of genes, sometimes in the vicinity of known oncogenes. In conjunction with selective pressure on transduced cells, this tendency can lead to insertional mutagenesis, the transformation of a somatic cell into a cancer cell. Lentivirus-based vectors have the advantage of being able to integrate into resting cells, allowing the transduction of stem cells that do not divide frequently, and also nerve cells. Additionally, they continue to be expressed long term, even in vivo, and they have a better safety profile as they do not prefer transcriptional start sites for integration.

Gene therapy clinical trials using retroviral vectors to treat X-linked severe combined immunodeficiency (X-linked SCID) represent a successful application of gene therapy to cure this monogenic disease [14]. The gene for the common gamma chain was transferred into hematopoietic stem cells of patients with X-linked SCID, which allowed the maturation of functional T cells. More than 20 patients have been treated in France and the United Kingdom. Unfortunately, four children in the French trial and one in the UK trial developed leukemia as a result of insertional mutagenesis by the retroviral vector. These children were treated for their leukemia and responded well, with the exception of one patient. The rest of the treated children did not develop leukemias and are considered cured with a normal immune system [15].

The leukemias in this particular clinical trial were caused by several occurrences: (1) the use of oncoretroviral vectors that integrated right next to an oncogene, the LMO2 gene. (2) The gene for the common gamma chain codes for a growth factor receptor and was always upregulated using a strong promoter, producing a highly active growth factor receptor. (3) There was enormous selective pressure on the very few transduced stem cells to produce functional T cells, selecting for the ones that grew the fastest. The application of a different vector, regulated growth factor genes, and safety measures such as suicide gene that would have allowed the ablation of transduced cells and their progeny would have made this gene therapy application safer.

Other gene therapy trials using retroviral vectors to treat SCID caused by adenosine deaminase (ADA) recently showed excellent success resulting in the cure of several children, without adverse events [16].

HIV gene therapy clinical trials were also conducted, with both retroviral and lentiviral vectors, without adverse events. However, their clinical efficacy remained low. These clinical trials will be discussed in detail in later chapters.

Other Viruses

Adenoviruses carry their genetic material as double-stranded DNA. In humans, they cause respiratory, intestinal, and eye infections and also the common cold. Adenoviruses bind to the target cell and transfer their DNA into the cell. Viral DNA is not integrated into the host cell's genome and remains relatively stable inside the nucleus of the host cell, producing viral messenger RNA utilizing the cell's machinery and quickly turning the cell into a virus factory. The DNA is not replicated during cell division and is only available transiently, as it gets degraded in the long run. As the adenovirus destroys the host cell relatively quickly, this is not a disadvantage in the life cycle of the virus. Rather, the virus depends on quick infection of other cells.

Adenoviruses can also be used as gene therapy vectors and can be made replication incompetent by deletions in the viral genome. These vectors are known for their efficient gene transfer into both resting and dividing cells. However, gene transfer is transient, and for prolonged gene expression, re-administration of the vector is required. Additionally, the viral particle itself has been shown to have toxicity, in the absence of viral replication. It can cause inflammation and immune reactions, which can be severe at high particle concentrations.

In spite of these disadvantages, the adenoviral vector system has been used frequently for direct in vivo administration in gene therapy clinical trials for cancer. An adenoviral p53-based gene therapy product was approved by the Chinese FDA in 2003 for treatment of head and neck cancer [17]. A similar gene therapy approach was not approved by the United States FDA in 2008.

Concerns about the safety of adenoviral vectors for in vivo administration were raised after the death of 18-year-old Jesse Gelsinger in 2000, who participated in a gene therapy clinical trial where high numbers of vector particles were infused intrahepatically. The death of the patient was also due, at least partially, to neglect of good clinical practice by the principal investigator [18, 19]. After this incidence, adenoviral vectors were heavily modified to improve safety, and clinical trial regulations were modified to be much more stringent.

Adeno-associated viruses (AAV) are part of the parvovirus family. They are small viruses carrying single-stranded DNA as their genetic information. Wild-type

AAV is unique as it can insert its DNA into chromosome 19, at a specific site. This directed integration of DNA is highly precise. Most people are carriers of AAV, which by itself is nonpathogenic as it needs adenovirus as a helper virus to replicate.

AAV can be used as a gene therapy vector as it can be engineered to be replication defective by not carrying viral genes essential for replication. AAV vector DNA, in which viral genes are substituted for the gene to be transferred, does not integrate into the host cell genome as it is lacking the information for this task. Instead, the recombinant vector genome fuses at its ends via the inverted terminal repeats (ITRs) to form DNA circles which persist episomally. AAV vector DNA is expressed for a prolonged period of time, most likely due to these episomal DNA circles.

The disadvantages of AAV vectors are the small amount of DNA they can carry and the difficulty of vector manufacturing. Only recently, clinical grade vector manufacturing has become more manageable. The great advantage of AAV is that it is nonpathogenic, can be administered in vivo without great risk, is non-integrating, can transduce resting cells such as nerve cells, and is expressed for a prolonged period of time. Several clinical gene therapy trials with AAV are ongoing, among them therapies for muscle and eye diseases [20]; however, clinical trials for neuro-degenerative diseases, where AAV vectors are used to deliver genes into the brain, are also in preparation.

Herpes Simplex Virus (HSV) Vectors: Wild-type herpes viruses (HSV-1 and HSV-2) are double-stranded DNA viruses with a relatively large genome encoding about 200 genes. The viral DNA is encapsulated in a protein capsid surrounded by a lipid bilayer. Herpes viruses engage specific cellular receptors for attachment to the target cell and enter the cell through membrane fusion. After entry into the cell, viral DNA is immediately transcribed, and "early proteins" are made which are needed for the regulation of virus production. "Late proteins" are transcribed after the viral DNA has migrated into the cell nucleus, forming the viral capsid and viral surface receptors. Herpes viruses tend to infect neural cells and can persist in them latently for prolonged periods of time. This may be mediated by herpes viral DNA binding to a specific neural protein involved in regulating viral latency. Initial HSV infection causes painful blisters in mucous membranes, but can usually be con-trolled by the immune system of the infected host organism. However, the virus can be reactivated and emerge from latency triggered by another viral infection such as the common cold, influenza, etc., causing the same painful symptoms. HSV infec-tion can become dangerous and even lethal in patients who are immunosuppressed. HSV can be used as a gene therapy vector with several advantages. The vector can carry a large DNA payload, it is neurotropic with strong transgene expression in neuronal cells, and latency can be exploited. This would make HSV-based vectors excellent candidates for gene transfer into neurons. HSV vectors can be engineered as attenuated replication-competent vectors, as replication-incompetent vectors with partial deletion of viral genes, and as "amplicons," which are defective helper-dependent vectors. HSV vectors have already been used in cancer gene therapy clinical trials [21].

Chapter 2
History of Gene Therapy

It is difficult to pinpoint the beginning of gene therapy, but 1967 can be considered as the beginning of at least the discussion about gene therapy. Marshall Nirenberg, who won the Nobel Prize in physiology in 1968, wrote in a 1967 paper about "programming cells with synthetic messages," recognized the usefulness of this procedure but also discussed its potential pitfalls and dangers [22].

Between 1970 and 1973, the American physician Stanfield Rogers collaborated with a German physician to develop a treatment for hyperargininemia. Two sisters suffering from this disease were chosen to be treated, in an experimental procedure, with Shope papilloma virus (SPV). Although Rogers called it a "wart virus," this virus can cause malignant transformation of cells in humans. Rogers believed that the virus would cause expression of the gene that regulated the production of arginine. This could later be demonstrated to be a false assumption. A final paper in 1975 reported the gene therapy experiment failed [23].

In the 1970s, research into recombinant DNA was moving forward. Extreme foresight has to be credited to the National Institutes of Health (NIH) in 1974 when it took the lead in regulating recombinant DNA research. A regulatory oversight body was created, which was called "the Recombinant DNA Advisory Committee (RAC) to the NIH Director," with RAC members initially being experts in mainly recombinant DNA technology. Over time membership was expanded to individuals coming from a wide range of scientific and medical disciplines, including ethicists and members of patient and other lay communities.

The RAC was initially put in charge of approving research projects involving recombinant DNA in NIH-funded laboratories in the United States. The RAC then got involved in regulating gene marking research projects and finally started to review gene therapy protocols together with the United States Food and Drug Administration (FDA). While the RAC would review the soundness and merit of the scientific aspect of the recombinant DNA technology applied, the FDA would focus on the safety and efficacy of the genetically modified products, including their manufacturing processes. Regulations that both the RAC and the FDA apply are based on the guidelines on human experimentation that stem from the work of the National

G. Bauer and J.S. Anderson, *Gene Therapy for HIV: From Inception to a Possible Cure*, SpringerBriefs in Biochemistry and Molecular Biology, DOI 10.1007/978-1-4939-0434-1_2, © Gerhard Bauer and Joseph S. Anderson 2014

Commission for the Protection of Human Subjects, as documented in the Belmont Report from 1978. Specific regulations were established that stipulated that recombinant DNA research proposals had to go through several review processes (US Office of Science and Technology Policy 1991). For a clinical trial involving recombinant DNA, approval was first required by the home institution's Institutional Biosafety Committee (IBC) and the Institutional Review Board (IRB); final approval was then required by the RAC. These regulations applied to all NIH-funded institutions involved in recombinant DNA research, even if the specific project in question was not funded through NIH money and even if the research was not taking place in the United States.

A first attempt of applying human gene therapy was conducted in 1980 under rather questionable circumstances by Martin Cline at the University of California, Los Angeles (UCLA). Without obtaining approval from the UCLA IRB and the other regulatory bodies, Cline performed a recombinant DNA transfer of the beta-thalassemia gene into bone marrow cells of two patients with beta-thalassemia, in Italy and Israel. At the time, IRBs had not yet been established in Italy; additionally, the principal investigator did not fully disclose the exact nature of the gene transfer clinical trial he was planning to conduct to the IRB in Israel. In October of 1980, the Los Angeles Times received information about these studies and published the details of Cline's recombinant DNA treatments [24]. Cline suffered grave consequences. He was forced to resign his department chairmanship and lost some grants, and for a period of 3 years, all of his applications for grant support were accompanied by a report of the investigations into his activities from 1979 to 1980.

In light of Dr. Cline's experiment, and at the prompting of the National Council of Churches, the Synagogue Council of America, and the United States Catholic Conference, the President's Commission for the Study of Ethical Problems in Medicine and Biomedical and Behavioral Research, a congressionally mandated group founded in 1978 and working independently from 1980 to 1983, became involved with the issue of gene therapy. The group released a landmark study called "Splicing Life" in 1982 [25] which defended the continuation of gene therapy research strongly. In this study, the laboratory risks associated with gene therapy research were reviewed carefully, it responded to concerns that scientists were "playing God," and it was concluded that a distinction can be made between acceptable and unacceptable consequences of gene therapy research. The President's Commission also suggested that the RAC should broaden the scope of its reviews to include ethical and social implications of gene therapy.

In 1984 the RAC created a new committee, called the Human Gene Therapy Working Group (later called the Human Gene Therapy Subcommittee (HGTS)), specifically to review clinical gene therapy protocols [26]. The first task of the Working Group was to produce a reference document, "Points to Consider for Protocols for the Transfer of Recombinant DNA into the Genome of Human Subjects." The intention of this document was to guide investigators applying for RAC approval of clinical gene therapy protocols [27].

In fall of 1985 the RAC Subcommittee had finished and published its "Points to Consider" document and was waiting to receive clinical gene therapy protocols for review. It took almost 3 years until the first protocol that was presented to the RAC in 1988 was a gene marking study by Steven Rosenberg. He proposed to use gene marking techniques to track trafficking of tumor-infiltrating blood cells in cancer patients. However, no actual "gene therapy" was proposed. After several months of discussion among the HGTS members and requests for additional information from the investigator, the protocol was finally approved in December of 1988 via mail ballot. The gene marking study was initially off to a bad start, since a lawsuit filed by the Foundation on Economic Trends questioning the validity of the review process halted it. Eventually, Rosenberg was allowed to continue and performed the gene marking in humans with fruitful results [28].

We will now mainly focus on the history of retroviral- and lentiviral-mediated clinical gene transfer studies. It should be pointed out, however, that many other gene therapy clinical trials utilizing nonviral vector gene transfer, adenoviral vector, and adeno-associated viral vectors have been carried out to this day with the majority of them testing novel gene therapy approaches for cancer, metabolic diseases, or hemophilia.

In 1990 the HGTS received two protocols to review. The first protocol was from Michael Blaese and W. French Anderson for T lymphocyte-directed gene therapy for ADA SCID. Severe combined immunodeficiency (SCID) caused by ADA deficiency is a monogenic disease, which means that just one gene in the human genome is defective; in this particular disease, it is the gene responsible for the production of the enzyme adenosine deaminase (ADA). ADA is needed as a detoxifying agent in the maturation process of T cells. If ADA is lacking, T cell function is severely impaired, leading to the absence of a cellular immune response, leaving patients vulnerable to recurrent opportunistic infections and even death from these infections. Initially it was thought that the gene should be inserted into autologous hematopoietic stem cells (HSCs) to improve upon allogeneic bone marrow transplantation which can cure ADA deficiency but is also associated with high morbidity and mortality in ADA patients. Disappointing gene transduction and engraftment results in nonhuman primates, however, convincing the researchers not to use autologous HSCs, since retroviral vectors could not transduce them efficiently. HSCs are mainly resting cells, and retroviral vectors can only integrate into dividing cells. It was decided that autologous peripheral blood T cells would be a better target, as they can be stimulated in culture to divide. It was possible to obtain enough peripheral T cells from ADA patients using an apheresis procedure. The HGTS reviewed the protocol, which was really the first human gene therapy protocol, and approved it. Two children with ADA SCID were then treated with T cell gene therapy for ADA SCID in 1990. Transduced autologous T cells were infused, and the children were followed for the survival of transduced T cells and any signs of clinical benefit. It could be demonstrated that transduced peripheral blood T cells do persist in vivo over years and produce ADA. However, the level of ADA in the patient remained too low to contribute to a clinical benefit. Both patients did not show any adverse reactions to the treatment [29].

The other clinical gene therapy protocol was received again from Steven Rosenberg. He wanted to use the same tumor-infiltrating blood cells he had previously investigated in his gene marking study as delivery vehicles for a tumor necrosis factor gene designed to kill tumor cells. His protocol was also approved [30].

At this point it should be mentioned that every clinical gene therapy study presented to the RAC for approval also has to be reviewed and approved by the local Institutional Review Board (IRB), by the Institutional Biosafety Committee (IBC), and by the FDA, which is the final authority for approval of such a study through an Investigational New Drug (IND) application.

In 1993 Andrew Gobea was born with ADA SCID; prior to his birth, genetic screening had already shown that he had SCID. By that time, research conducted by Donald Kohn and Jan Nolta at Children's Hospital Los Angeles suggested the possibility that novel HSC culture methods allowed for normally resting HSCs to be driven into cell cycle. This made gene transduction with retroviral vectors a possibility. A clinical trial for stem cell gene therapy for ADA SCID was presented to the RAC, to the FDA, and to the other regulatory agencies and was approved. Umbilical cord blood, which contains hematopoietic stem cells, was collected from Andrew's umbilical cord immediately after birth. The HSCs were selected using the first clinical grade CD34+ selecting device, made by CellPro, a company which has since gone out of business over the CD34+ antibody patent fight. Selected CD34+ cells were cultured and transduced with a retroviral vector transferring the ADA gene. After transduction, the cultured cells were washed and infused into the patient. Gene marking in the peripheral blood T cells appeared and was stable for years. Three other children with ADA SCID were also treated with the same stem cell gene therapy approach, and similar results were obtained. As standard of care, all children that had undergone stem cell gene therapy were still kept on injectable PEG-ADA enzyme to maintain functional T cells and to prevent opportunistic infections. However, after stable gene expression, although low, could be measured consistently, withdrawal of injectable PEG-ADA was tried. Gene marking and ADA expression in the peripheral blood increased, as a selective survival advantage allowed for the transduced, gene-expressing cells to expand, but the numbers of gene-expressing T cells did not reach therapeutic levels. The children had to be put back on injectable PEG-ADA. In spite of not achieving the wanted clinical outcome, this first stem cell gene therapy study still paved the way for all other stem cell gene therapies to follow [31].

Similar results were obtained in initial attempts to correct chronic granulomatous disease and Gaucher disease. The therapeutic genes were again inserted into hematopoietic stem cells; however, those genes did not confer a selective advantage upon gene-transduced cells. In a Phase I clinical trial for X-linked chronic granulomatous disease (X-CGD), the investigators added flt3-ligand and granulocyte-macrophage colony-stimulating factor into the transduction culture medium and utilized fibronectin as a cell attachment matrix in their 4-day transduction protocol. The patients underwent multiple cycles of mobilization and infusion of transduced cells without preconditioning. Nine months after transplantation, the levels of gene-corrected neutrophils were only 0.06–0.2 %, which was well below the desired 5–10 % required for therapeutic effects [32]. In patients with Gaucher disease, there was also very low gene marking, little or no gene expression, and no clinical benefit [33].

In the early to mid-1990s, gene marking trials were carried out in patients undergoing autologous HSC transplantation for cancer. These trials had a dual purpose: first, to investigate the source of cancer relapse and, second, to explore gene transfer efficiency into hematopoietic stem cells and gene marking persistence. Even with full bone marrow ablation, patients showed levels of marked blood cells in the periphery well below 1 %. Sometimes there was not even any gene marking [34–36]. However, taken together these studies were able to demonstrate that autologous mobilized peripheral blood CD34+ cells contained long-term repopulating stem cells which could be transduced, to some degree, with retroviral vectors. They also clearly demonstrated that better gene therapy vectors and better transduction conditions for HSCs had to be developed.

In 1997, the first child in the world was treated with stem cell gene therapy for HIV at Children's Hospital Los Angeles. An HIV-1 RRE decoy gene was transferred into CD34+ cells from the bone marrow of HIV-1-infected pediatric patients in a Phase I feasibility and safety study to evaluate potential adverse effects from such a gene transfer procedure. Feasibility was defined as the ability to obtain an adequately large bone marrow aspirate from which enough CD34+ cells could be isolated using a closed system magnetic cell separator. It was estimated that at a minimum, $1 \times 10e6$ CD34+ cells per kg body weight would be needed. Efficacy was assessed by determining whether the cultured, gene-transduced bone marrow cells would engraft and produce white cells in the peripheral blood expressing the transferred anti-HIV gene. To test the hypothesis that anti-HIV-1 gene-expressing cells would have a selective survival advantage in a patient with HIV infection, a comparative marking approach was used. Half of the cells were transduced with a control vector transferring neomycin resistance, while the other half was transduced with a vector transferring the anti-HIV gene in conjunction with neomycin resistance. It should be pointed out that at that time, the transfer of neomycin resistance was allowed by the regulatory bodies and could be used as a marker for the detection of transduced cells. If there were a selective survival advantage of anti-HIV gene-expressing peripheral cells, these would accumulate preferentially as compared to neomycin gene-expressing cells. Sufficient numbers of CD34+ cells from the bone marrow of this child were isolated successfully and could be transduced with the retroviral vector. The transduced CD34+ cells were administered to the child without preconditioning. Gene marking in the peripheral blood developed over the following weeks and could be detected at low levels, which was expected as no preconditioning of the bone marrow was used. However, gene marking declined significantly and disappeared after several months. A selective survival advantage of the gene-transduced cells in the face of a viral load did not develop, since the patient had to take antiretroviral therapy (ART) and had no viral load, as this was demanded by the FDA due to ethical reasons [37]. This and other stem cell gene therapy clinical trials for HIV will be discussed in greater depths in the following chapters.

In 1998, the first adult was treated with stem cell gene therapy for HIV at City of Hope National Medical Center. A ribozyme directed against the tat/rev region of the HIV transcript was used as the anti-HIV gene. CD34+ cells were collected through apheresis and a CD34+ cell isolation was performed. In this Phase I clinical trial, five adult patients were enrolled. No marrow conditioning was applied. Low gene marking in the peripheral blood was detected and persisted over several months [38].

In 1999, the first marrow-ablated adult with HIV-induced B cell lymphoma was treated with stem cell gene therapy for HIV and autologous bone marrow stem cell transplantation. The same anti-HIV gene was applied as in the previous City of Hope clinical trial. After cell infusion, high gene marking could be detected in the peripheral blood, with excellent expression of the anti-HIV gene in the transduced peripheral blood cells. However, gene marking decreased and almost completely disappeared after several months. In neither clinical trial, withdrawal of ART was allowed; therefore, a selective survival advantage of anti-HIV gene-expressing cells could be demonstrated [38].

Also in 1999, another pediatric clinical trial of stem cell gene therapy for HIV was conducted at Children's Hospital Los Angeles. Children with a low HIV viral load while on ART were the target population. The anti-HIV gene was a transdominant negative Rev protein, the best anti-HIV gene at that time. One child was treated with gene-transduced CD34+ cells isolated from a bone marrow aspirate. No marrow conditioning was used. Low-level gene marking developed in the peripheral blood after several weeks and again disappeared after a few months. The patient, however, stopped the antiretroviral medication at around 12 months after cell infusion, due to side effects of the medication. The patient was still followed at Children's Hospital for the gene therapy study. Interestingly, the HIV viral load had increased significantly, and gene marked peripheral blood T cells again appeared. The numbers of gene marked peripheral blood cells correlated with the increase in viral load. This demonstrated, although only anecdotally, as this was a single patient, the predicted selective survival advantage of anti-HIV gene-expressing HIV target cells in the face of a viral load [39].

In October 1999, the death of 18-year-old Jesse Gelsinger was reported [40]. Gelsinger had participated in a gene therapy clinical trial using adenoviral vector for the treatment of ornithine transcarbamylase (OTC) deficiency. A very high dose of adenoviral vector that was known to the PI having caused fatal adverse events in a primate study was administered to the patient intrahepatically. Gelsinger died 4 days after the infusion from a massive immune reaction. Gelsinger was not a patient that absolutely needed gene therapy since his condition was not as severe and could be treated by diet restriction and medication. It was the PI who persuaded Gelsinger to participate. Clearly, the PI had neglected to inform the FDA, the RAC, and all other regulatory bodies about the adverse reactions in the primate experiment and had neglected good clinical practice by persuading Gelsinger to participate in the clinical trial. Gelsinger's death therefore raised questions about researcher entrepreneurial activities and conflict of interest and about government oversight procedures. The United States Senate held hearings on this topic on February 2, 2000, and the FDA sent out the "March 6 letter" to all investigators conducting gene therapy clinical trials announcing new and heightened scrutiny in gene therapy oversight. This has resulted in increased reporting of adverse effects and renewed control by both the NIH RAC and the FDA. Gelsinger's death also resulted in federal charges being brought against the principal investigator and the university who conducted this trial [41]. Finally, a settlement with the US Office of the Attorney General was reached in February of 2005.

The success of a multicenter clinical trial of stem cell gene therapy for children with X-linked SCID conducted from 2000 to 2002 in France was put into question when two of the ten children treated at the site in Paris developed a leukemia-like condition caused by clonal expansion of gene-expressing T cells. All clinical trials of stem cell gene therapy were halted but were allowed to continue after re-reviews of the affected gene therapy protocols in the United States, the United Kingdom, France, Italy, and Germany [42].

In 2007, a death in a gene therapy trial for arthritis was investigated and found to be caused by factors not related to the gene therapy clinical application. This case prompted discussion of including procedures for investigating other illnesses that might occur during a gene therapy study in the clinical trial protocol [43].

Chapter 3
Principles of HIV Gene Therapy

The human immunodeficiency virus (HIV) belongs into the family of retroviruses and within this virus family into the group of lentiviruses; among the hallmark characteristics of lentiviruses are long incubation periods and persistent infection. These viruses are single-stranded RNA viruses which carry their own enzymes that they use to reverse transcribe their RNA into double-stranded DNA, which they then permanently integrate into the host cell genome. Once inside the host body, HIV persists lifelong and if not treated causes the acquired immunodeficiency syndrome (AIDS), which is an immunodeficiency that is mediated by the loss of CD4-positive T helper cells, eventually leading to complete destruction of the human immune system. This makes the patient vulnerable to opportunistic infections, which after prolonged illness invariably result in death.

HIV continues to affect millions of people around the world. There is no effective vaccine available, and current antiretroviral treatments are only effective at suppressing viral load but do not cure HIV infection, as its genetic information is permanently integrated into the genome of long-lived host cells [44]. The high mutation rate of HIV has made it difficult to design drugs against the virus. The viral reverse transcriptase is inherently failure prone, and about 1 in 1,700 nucleotides it incorporates into newly synthesized DNA is incorrect [45]. This leads to many new viruses with errors in their genetic information. Many of these errors cause no change in viral phenotype or fitness, some make the new viruses less fit, and some give the newly arising viruses unique features allowing them to adapt to antiviral drugs very quickly, as selective pressure is applied by drugs to only allow those viruses to replicate that do not get inhibited by these drugs.

For these particular reasons, antiretroviral monotherapy does not work, which was unfortunately seen in the 1980s after the first anti-HIV drug, AZT, was widely applied. Only a few weeks into the drug treatment, resistant viruses emerged, and AZT lost its therapeutic effect [46]. The only way to overcome this issue of emergence of drug resistance in HIV infection is to completely and permanently suppress HIV replication, so no reverse transcription can occur, effectively eliminating the viral mutation ability, and to attack multiple targets within the HIV life cycle,

G. Bauer and J.S. Anderson, *Gene Therapy for HIV: From Inception to a Possible Cure*, SpringerBriefs in Biochemistry and Molecular Biology, DOI 10.1007/978-1-4939-0434-1_3, © Gerhard Bauer and Joseph S. Anderson 2014

so it becomes much more complicated for the virus to produce mutations that are resistant to all of these multiple target therapeutics. Since 1996, antiretroviral drug combinations have therefore been used to treat HIV infection [46, 47]. This strategy has been extremely successful as viral loads can usually be dropped to undetectable levels and patients do not progress to the state of AIDS anymore. However, lifelong drug treatment is a huge problem, as it is associated with lifelong high cost (about $35,000 per year per patient); also long-term toxicities, particularly in an aging population, are a concern. To treat HIV infection successfully, very high compliance in taking the combination of drugs is required. If the drugs are not taken regularly, due to different half-lives of the individual drugs in the patient, HIV strains resistant to the individual drugs used in the "drug cocktail" can arise.

Although during the beginning of successful drug treatment of HIV-infected individuals attempts were made to completely eliminate HIV from a patient, these attempts have all failed. HIV persists in latent reservoirs, in so-called sanctuary sites, suspected to be long-lived cells such as memory T cells, lymphocytes and monocytes in the gut mucosa, dendritic cells, etc. Whenever antiretroviral therapy (ART) is withdrawn, HIV rebounds from these latent reservoirs. The exact mechanism of latency of HIV is still under investigation [48, 49].

Clearly, novel methods for HIV treatment with the possibility of a cure are needed. Gene therapy for HIV is such a novel method, although it has been conceived many years ago. In 1988 David Baltimore in an article in "Nature" coined the phrase "intracellular immunization," which means making cells resistant from the inside (by application of genetic manipulation) against the detrimental effects of HIV [50]. Several groups of researchers, among them ours, took this concept very seriously and have pursued it to this day, with remarkable improvements of the process. We also believe that stem cell gene therapy offers a possible path to a cure for HIV.

In order to understand the principle behind HIV gene therapy, it is necessary to have a closer look at the HIV life cycle. HIV infects the CD4+ cells of the human immune system by binding to the CD4 receptor as its main receptor and also to a secondary receptor, the CCR5 coreceptor (macrophage tropic HIV strains) or the CXCR4 coreceptor (T cell tropic HIV strains) on the host cell through interactions with glycoproteins, carried by the HIV viral envelope (Fig. 3.1). Envelope–receptor binding allows the viral particle to fuse with the host cell membrane and enter the cytoplasm. Once inside the host cell, HIV "uncoats," releasing its genetic information from the viral core that has been used to protect the viral RNA. Within this viral core, a "pre-integration complex" has formed. Among other components, the HIV pre-integration complex includes the complete viral RNA genome and the unique viral enzymes reverse transcriptase and integrase. Reverse transcriptase converts the RNA genome into double-stranded DNA. Viral capsid protein, which is also part of the pre-integration complex, then attaches to a cellular transport mechanism which actively transports the newly synthesized viral DNA into the nucleus. The viral integrase then pseudorandomly integrates the double-stranded, full-length viral DNA into the genome of the target cell (Fig. 3.1) 1, 2, 3.

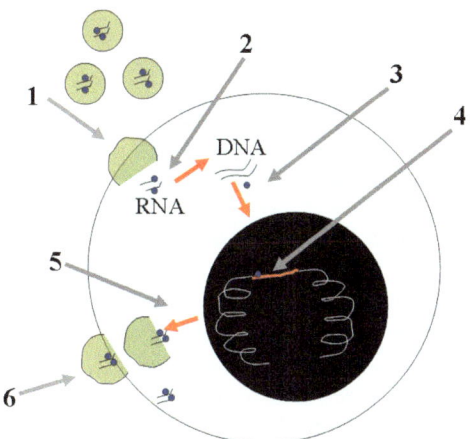

Fig. 3.1 HIV life cycle. (**1**) Attachment of HIV particle to the target cell mediated by envelope glycoproteins which bind to the CD4, CCR5, or CXCR4 receptors, subsequent fusion with cell membrane. (**2**) Reverse transcription of viral RNA into double-stranded DNA by viral reverse transcriptase. (**3**) Transport of newly synthesized DNA into host cell nucleus by cellular transport mechanism. (**4**) Viral integrase-mediated HIV DNA integration into the host cell DNA. (**5**) Transcription of integrated DNA into spliced short and unspliced long viral RNAs, transport of these into cytoplasm, and translation into viral proteins cleaved to length by viral protease. (**6**) HIV particle assembly and RNA packaging, budding of the infectious particle from the host cell

After integration, HIV immediately upregulates its genetic information due to the action of several regulatory genes encoded by the virus and initiates transcription of short RNAs which are spliced viral RNAs. These short viral RNAs are translated into viral regulatory proteins that induce an environment in the host cell suited for upregulation of viral replication and viral particle synthesis. This enables infected cells to start producing a large amount of viral particles (Fig. 3.2), leading to a spike of viral load in the infected individual. This viral spike can only be kept under control by a vigorous immune response mounted by the host immune system. This initial immune response is also correlated with clinical symptoms such as fever, rash, and flu-like symptoms called "seroconversion illness." Only after the immune system has been able to eliminate most of the circulating virus by killing many of the infected cells responsible for a high viral output will a "steady state" be achieved that determines the set point of the viral load in the patient [51]. A long-lived "reservoir" of HIV-infected cells will also be created at that time, which is achieved by viral integration into cells with low cell cycle turnover.

In the past, a so-called "latent" state of viral infection over many years was proposed. However, HIV is not really latent in most infected individuals, but HIV replication is rather kept somewhat under control by the immune system trying to eliminate cells that replicate HIV. If untreated, this state can last for 8–10 years. During this time, HIV has enough time to infect many CD4+ T cell clones and is able to eliminate in the long run also those clones that are responsible for keeping

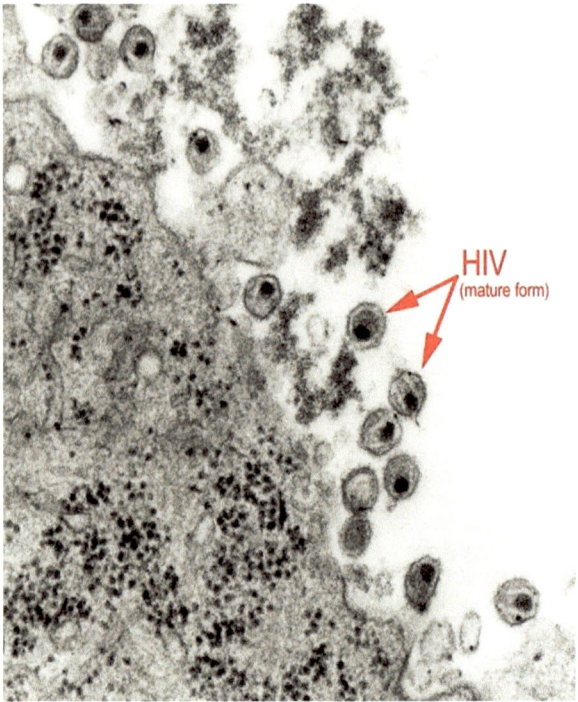

Fig. 3.2 Transmission electron microscope photograph of HIV budding from an infected cell. After integration into the host cell DNA, HIV immediately upregulates expression of viral RNAs and protein synthesis, creating a large burst of new particles. This viral particle burst is shown here. HIV can be recognized as a small, round particle with a dark core, the capsid, which contains the viral RNA

HIV under control. At this stage, HIV replication increases dramatically, destroys the remaining CD4+ T cells, and leads to the clinical manifestations of AIDS.

ART treatment for HIV consists of a combination of drugs that target different stages of the lentiviral life cycle which present multiple obstacles for viral replication. Common drugs used in ART are the following: (1) nucleoside and nucleotide reverse transcriptase inhibitors (NRTI) which inhibit reverse transcription by incorporation into the newly synthesized viral DNA strand causing strand breaks; (2) non-nucleoside reverse transcriptase inhibitors (NNRTI) which inhibit reverse transcription by inhibition of primer as these drugs can bind to the active region of the enzyme in a highly specific manner; (3) protease inhibitors which inhibit the function of another unique viral enzyme responsible for cleaving long viral protein strands needed for particle formation; (4) integrase inhibitors which inhibit the viral integrase sterically, preventing viral DNA integration; and (5) entry inhibitors, which prevent binding of the viral particle to the secondary receptor, or fusion inhibitor, which prevent fusion and entry of the viral particle. Only using such a multitude of drugs, it is nowadays possible to find the right regimen for a patient which allows

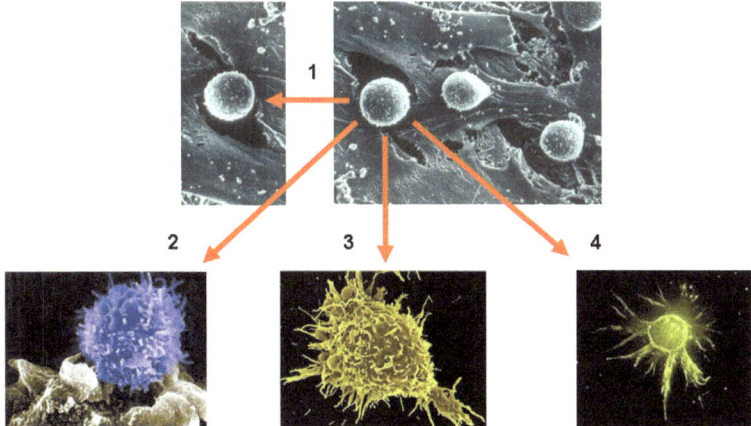

Fig. 3.3 HIV target cells derived from hematopoietic stem cells. Hematopoietic stem cells (HSCs) give rise to all blood cells while maintaining, at the same time, the HSC pool. This is accomplished by asymmetric division: After cell division, one cell of the pair undergoes lineage differentiation into blood cell progenitors and finally mature blood cells, while the other cell of the pair remains a stem cell (**1**). HIV target cells are T cells (**2**), macrophages (**3**), and dendritic cells (**4**)

suppression of most HIV variants so they cannot sufficiently replicate, mutate, evade the drug treatment, and continue to destroy the host immune system [52].

Gene therapy for HIV can be conceived as a process of introduction of a gene or multiple genes into the genome of HIV target cells that, when active, will produce gene products with anti-HIV activity that can protect HIV target cells permanently from HIV infection and replication. Target cells for HIV are CD4+ T cells, macrophages, monocytes, dendritic cells, and even brain microglia. One very interesting aspect of all of these cell types is the fact that they are all of hematopoietic origin; that means they are derived from bone marrow stem cells (Fig. 3.3).

The main target cell for HIV, which is also associated with the pathology of the disease, is the CD4+ T cell present in the peripheral blood and in the lymphatic system. When the number of CD4+ T cell is diminished, immune function is also reduced. The CD4+ "T helper cell" is the pivotal cell within the immune system, responsible for eliciting clonal expansion of appropriate T cell fractions and also responsible for signaling to B cells for appropriate antibody production during an immune response.

It is feasible, for HIV gene therapy, in its simplest application, to insert anti-HIV genes into the primary target cells for HIV, peripheral blood T cells. These cells can be collected easily from a patient in an apheresis procedure, cultured and expanded in the laboratory, transduced with vectors carrying anti-HIV genes, and then reinfused into the patient. If enough T cells express anti-HIV genes, viral replication will be diminished, similar to the action of anti-HIV drugs. Due to its ease, T cell gene therapy has been applied in clinical trials of HIV gene therapy many times. Both retroviral and lentiviral vectors have been used for transduction, and a several anti-HIV gene strategies have been applied. However, there are great disadvantages to this treatment: There are billions of T cells, which can be potential targets for HIV in the peripheral blood and the lymphoid system, the majority of these cells not accessible to apheresis. In spite of repeated harvesting, culture, and transduction procedures, this HIV gene therapy approach will only make a small number of T cells in the body resistant to HIV, while the majority of T cells will remain easy targets for HIV. Additionally, T cells (except memory T cells) are not long lived, and T cell gene therapy will have to be repeated, if long-term therapeutic effects are desired. It has therefore been argued correctly that T cell gene therapy for HIV would be similar to ART and would need to be a lifelong therapy.

Is there a place for T cell gene therapy? It has been suggested that this type of HIV gene therapy could be used to augment ART if a patient were resistant to the drugs in the regimen. Indeed, the first HIV gene therapy clinical trial applying a lentiviral vector was based on this idea, and patients were treated that were resistant to ART [53].

Remembering the notion that all HIV target cells are derived from bone marrow stem cells, a different and more permanent HIV gene therapy approach can be developed. If anti-HIV genes are permanently inserted into the genetic information of bone marrow stem cells, all arising cells of the hematopoietic lineage will also carry these anti-HIV genes, as they will be passed on from the hematopoietic stem cells to their progeny during lineage differentiation. Finally, these anti-HIV genes will be present in mature blood cells, such as macrophages, dendritic cells, and T cells. As bone marrow stem cells are an inexhaustible reservoir for new peripheral blood cells, an inexhaustible source of HIV-resistant cells can therefore be created by transduction of bone marrow stem cells with anti-HIV genes.

Bone marrow stem cells can be harvested via bone marrow aspiration, or much easier, by collection of mobilized of peripheral blood stem cells through apheresis. Bone marrow stem cells can be found among the fraction of CD34+ cells, which can be isolated from the bone marrow or an apheresis product via immunomagnetic bead isolation. As compared to peripheral blood T cells, only a rather small numbers of CD34+ cells are needed, as robust engraftment of such isolated cells could be demonstrated in many clinical bone marrow transplantation procedures already at $2 \times 10e6$ CD34+ cells per kg body weight. If all CD34+ cells in the transduction culture could be transduced with anti-HIV genes, at least the theoretical possibility of replacing all target cells of HIV with HIV-resistant cells exists. Isolated bone marrow stem cells were transduced, in the past, with retroviral vectors transferring single anti-HIV genes. Currently, lentiviral vectors that transfer more than one anti-HIV gene are the state of the art. As learned from small molecule anti-HIV drugs, for an effective defense against HIV, it is important to simultaneously target multiple stages in the HIV life cycle.

This statement is just as important and true for gene therapy; therefore, multiple anti-HIV genes should be applied. Each of these anti-HIV genes should interfere with the HIV life cycle at a different stage, preferably pre-reverse transcription, so HIV has no chance of reverse transcribing and integrating. This makes mutations less likely and also prevents spread of HIV by simple cell division, when HIV is passed onto daughter cells by DNA replication.

Anti-HIV genes can be designed to interfere with many different steps in the life cycle of HIV. They should be very specific in their target as not to interfere with normal cellular functions and should be strong inhibitors of HIV, ideally shutting down HIV replication so viral mutations cannot arise easily.

Among the earliest anti-HIV genes were "antisense" RNA oligonucleotides, synthesized strands of RNA exactly complementary to portions of the HIV mRNA, which acts as the "sense" RNA. After annealing of sense and antisense RNA, translation of the messenger RNA is inhibited, and RNA degradation sets in. Many different kinds of anti-HIV antisense strategies for HIV were tried; most of them unfortunately had low efficiencies in inhibiting HIV replication. One antisense strategy, however, was markedly improved and made it into clinical trials. This antisense RNA was a long oligonucleotide directed against the mRNA of the HIV viral envelope. Interestingly, the vector transferring the message for this RNA was a first-generation lentiviral vector that could be mobilized from transduced cells. This vector was applied in T cell gene therapy for HIV [53].

Another anti-HIV gene strategy is to interfere with HIV-specific regulatory factors. For instance, the rev-responsive element (RRE) on the mRNA transcript of HIV is needed to export the long HIV mRNA from the cell nucleus into the cytoplasm where translation can be carried out efficiently. The Rev protein, produced early in the life cycle of HIV by a spliced HIV mRNA, binds to the rev-responsive element and facilitates transport of the long mRNA transcript. If Rev cannot bind, the mRNA cannot be transported out of the nucleus and HIV replication is inhibited. An RNA molecule called "RRE decoy" can be constructed and transferred by a retroviral vector. If enough of this RRE decoy can be made, all or most of the available Rev protein is bound up and not available for binding to the HIV mRNA. This principle was very successful in preclinical experiments using a retroviral vector which was then applied in the first pediatric stem cell gene therapy clinical trial for HIV [37].

Another anti-HIV strategy directed against the RRE is the transdominant negative Rev protein. This is a mutated Rev protein that can still bind to the RRE, but does not connect to the cellular transport mechanism that is needed to transport the long transcript of the HIV mRNA out of the cell nucleus. This strategy was extremely successful in preclinical experiments and was utilized in two clinical trials, one mediating gene transfer by gold particles and a gene gun [6] and the other one using a retroviral vector [39]. It should be pointed out that this was a protein gene therapy approach, as compared to an RNA gene therapy approach in the first two clinical applications. RNA approaches are advantageous as they will not elicit immune responses in the host. Protein approaches, however, allow for many more active gene products to be produced from one messenger RNA and can be more efficacious; however, they can potentially elicit an immune response.

The TAR decoy is an anti-HIV gene approach to inhibit transcriptional upregulation of HIV. Tat is an important regulatory factor for HIV expression and needs to bind to TAR, the transactivation response element on the HIV DNA. If Tat is bound by a small RNA decoy, Tat binding to TAR is inhibited, and potent inhibition of HIV replication is achieved, as transcription of the long HIV mRNA cannot be initiated. This could be demonstrated very successfully in preclinical experiments [54].

A completely different approach is the application of a ribozyme against the tat/rev transcript in the HIV mRNA. Ribozymes are catalytic RNA molecules that can be designed to specifically cut an RNA strand at a very defined target site. This was a promising strategy in the mid-1990s and very successful in preclinical experiments [55]. The ribozyme against the tat/rev mRNA was applied in two clinical trials using retroviral vectors.

The strategy above can even be improved upon if the ribozyme is exchanged for a short interfering RNA (siRNA). siRNAs or short hairpin RNAs (shRNAs) having the same function are short double-stranded RNAs with 20–25 nucleotides that can be constructed to induce "RNA interference," silencing of specific gene expression. siRNAs have been directed against the tat/rev part of the HIV mRNA, and potent silencing of HIV expression and replication could be observed in preclinical experiments [56].

All of the anti-HIV strategies described above target HIV after entry into the target cell and after integration. HIV therefore can be passed on, in its integrated form through cell division. In order to protect a cell from HIV integration, HIV inhibition prior to reverse transcription should be the preferred strategy. Among the first anti-HIV molecules accomplishing this was a gp41 fusion inhibitor, a target cell membrane-anchored c-peptide that inhibits fusion of the virus with the target cell membrane. This fusion inhibitor was tested in a clinical trial of HIV T cell gene therapy in Germany [57].

Another strategy to inhibit HIV prior to reverse transcription is the knockdown of the secondary receptor for HIV. There are two secondary receptors for HIV-1 on the target cells: the CCR5 receptor and the CXCR4 receptor. In their normal function, they act as chemokine receptors during an immune response. There are two major strains of HIV: macrophage tropic strains and T cell tropic strains. Their tropism is dependent on their ability to bind to the specific secondary receptor; macrophage tropic strains bind to CCR5, and T cell tropic strains bind to CXCR4. Interestingly, naturally arising resistance to HIV infection is mediated by a deletion of the CCR5 receptor. Individuals with a homozygous deletion of this receptor do not display any phenotype associated with this deletion. This is due to the fact that chemokine receptors have redundant function, which means that other chemokine receptors can make up for the lost function of the CCR5 receptor. If a ribozyme or siRNA is designed against the message for the CCR5 receptor, the receptor expression may be downregulated enough to prevent binding, and entry of HIV into the target cell may be inhibited. This hypothesis could be verified in multiple preclinical experiments [58].

A very interesting strategy preventing reverse transcription is viral uncoating inhibition. In old-world monkeys, a molecule called Trim5alpha has developed this function as a natural resistance strategy against retroviruses. Humans also have a Trim5alpha molecule; however, HIV has found a strategy to mutate around its function. Therefore, if such a molecule should be used as an anti-HIV strategy, Trim5alpha needs to be engineered to inhibit HIV from uncoating. Such a molecule is discussed in other chapters in this book [59].

The best and currently most sophisticated HIV gene therapy strategy is the combination of several anti-HIV genes into one gene therapy vector. The group at City of Hope has developed a triple combination lentiviral vector containing a ribozyme against the CCR5 receptor, an shRNA against the tat/rev transcript of HIV, and a TAR decoy, inhibiting transcriptional upregulation of HIV. This vector has already been used in a stem cell gene therapy clinical trial for HIV [60].

Another novel triple combination anti-HIV gene lentiviral vector was developed by our group, containing an shRNA against the CCR5 receptor, a modified Trim5alpha molecule inhibiting uncoating of HIV, and a TAR decoy [61]. This vector is discussed in more detail in other chapters.

Chapter 4
Gene Therapy Vectors

Gene therapy vectors derived from viruses of the Retroviridae family, including oncoretroviral, Moloney murine leukemia virus, and HIV-based lentiviral vectors, are important tools for long-term genome modifications which are required for HIV stem cell gene therapy. This is due to their ability to integrate into the genome of transduced cells. There are, however, many differences between retroviral and lentiviral vectors including levels of transduction efficiencies into hematopoietic stem cells, vector integration sites, and levels of transgene expression.

Both retroviral and lentiviral vectors reverse transcribe their RNA gene sequences into DNA which is then integrated into the host genome. Prior to integration, the vector pre-integration complex which contains the reverse-transcribed DNA containing the therapeutic gene(s) must traverse the nuclear membrane. Retroviral vectors require cell division for successful nuclear localization of the pre-integration complex from the cytoplasm to the nucleus and also for integration. Thus, retroviral vectors can only be used at high efficiency in specific cells which divide more readily but are not useful with quiescent cells. Lentiviral vectors, however, are capable of more efficient nuclear localization due to the incorporation of viral proteins (matrix p17) into the vector particle which are then used for nuclear transport. This trait is unique to lentiviral vectors and provides these vectors with the ability to integrate into both nondividing and dividing cells [62].

Retroviral vectors integrate near transcriptional start sites of active genes and on occasion can result in the activation of nearby oncogenes. Patients who were enrolled in a gene therapy clinical trial for X-linked severe combined immunodeficiency developed leukemia due to the select integration site of the therapeutic retroviral vector near the LMO2 oncogene [63]. HIV-based lentiviral vectors, however, also integrate in transcriptionally active regions, however not in transcriptional start regions [64]. As different cell types have different genes being transcribed, they bias towards integration sites that have been found to be cell-type specific.

Silencing of transgene expression is another problem associated with the use of retroviral vectors. Both the integration site and the levels of therapeutic transgene expression can be altered by the transduced cell's nearby chromatin. This is

G. Bauer and J.S. Anderson, *Gene Therapy for HIV: From Inception to a Possible Cure*, SpringerBriefs in Biochemistry and Molecular Biology, DOI 10.1007/978-1-4939-0434-1_4, © Gerhard Bauer and Joseph S. Anderson 2014

mediated by the fact that retroviral vectors prefer to integrate near transcriptional start sites of the host's genes. Lentiviral vectors are less prone to transgene silencing, again, due to their different integration patterns.

The first vectors used in HIV stem cell gene therapy trials were MLV-based retroviral vectors. Transgene silencing was a major issue, and loss of anti-HIV gene expression became problematic. However, due to HIV/AIDS being a lifelong chronic disease, continued anti-HIV gene expression is necessary to continue blocking further HIV replication. Due to the need for prolonged genetic modification and anti-HIV transgene expression, HIV-based lentiviral vectors have become more widely used today.

Lentiviral Vectors

HIV/AIDS is a chronic infection/disease, and constitutive and stable, lifelong expression of anti-HIV genes is therefore necessary to control further HIV replication. Therefore, lentiviral vectors, particularly HIV-based vectors, are of great interest due to their ability to transduce nondividing cells, including hematopoietic stem cells. They are also less prone to transgene silencing as compared to retroviral vectors. HIV-based lentiviral vectors are able to transduce both nondividing and dividing cells; however, cells in G0 phase are more difficult to transduce than dividing cells.

The earliest HIV-based lentiviral vectors utilized for gene therapy studies were capable of replicating since most of the HIV proteins remained in the vector. The only insertion of the first HIV-based lentiviral vector was that of the chloramphenicol acetyltransferase gene which took the place of the HIV *nef* gene. Safety modifications, however, were then proposed to make the vector's replication incompetent. The first attempts were made by separating the viral genes into two separate plasmids, one encoding the HIV DNA with a deletion of the envelope gene and the second plasmid encoding the envelope gene. This would allow for only one round of transduction of the therapeutic gene into the target cell, but no replication of the viral particle due to the deletion of the envelope gene in the integrated vector. This also allowed for the therapeutic gene of interest to be inserted into the envelope region or in the nef region, as nef is not required for lentiviral vector production.

The "first-generation" HIV-based lentiviral vectors were developed with greatly increased safety by separating the critical vector production genes into separate plasmids. This was mainly due to the concern of generating replication-competent lentiviruses by recombination between vector sequences and/or infective HIV sequences. This also eliminated the possibility of HIV proteins from being expressed in transduced cells. Only the therapeutic genes would be expressed. The "first-generation" HIV-based replication defective lentiviral vectors were generated by transfection of three separate plasmids: (1) a vector packaging construct, (2) a vector envelope construct, and (3) the transfer vector which encodes the therapeutic gene of interest [65]. The packaging construct encodes the HIV genes required for encapsidating the vector genome and reverse transcribing/integrating the vector DNA. The envelope construct encodes a viral glycoprotein for attachment and fusion to targeted cells for transduction. The third plasmid encodes many *cis*-acting viral sequences needed for assembly, packaging, and integration of the vector. These include the viral long terminal repeats for integration, the psi signal for genome packaging, and the rev response element for nuclear to cytoplasmic transport of the transcribed vector RNA for proper translation of the therapeutic gene(s). The transfer vector does not contain any sequences for HIV proteins. Rather, transgene expression occurs from internal promoters incorporated with the therapeutic transgene and not from the viral long terminal repeats. These can include polymerase-II promoters for genes which need to be translated or polymerase-III small RNA promoters including H1 and U6 for driving the expression of therapeutic RNAs.

Considerations of target cells for transduction were quick to realize that utilizing the HIV envelope gp160 would restrict transduction to CD4+ cells. Therefore, HIV-based vectors were pseudotyped with other viral envelopes, particularly the vesicular stomatitis virus glycoprotein (VSV-G). This envelope protein is pantropic, meaning that it has a broad range of cell types that it can bind/fuse to because of a ubiquitous cell membrane protein which it attaches to. Therefore, by pseudotyping HIV-based lentiviral vectors with VSV-G, a wide array of cell types can be transduced, including human hematopoietic stem cells.

Further modifications were made to the HIV-based lentiviral vectors which were termed "second generation." These modifications consisted of deleting most of the viral genes including the HIV accessory proteins from the vector production system. These included the HIV virulence genes, *vif*, *vpr*, and *vpu* which were deleted along with *nef* as they are not required for vector production [66]. The long terminal repeats located in the transfer vector and used for vector integration were also modified to abrogate transcriptional activation and to eliminate the occurrence of recombination events with wild-type HIV which might generate replication-competent vector particles. This was accomplished through the generation of self-inactivating lentiviral vectors. Normally, the HIV long terminal repeat contains a U3, an R, and a U5 region. The U3 sequence is the critical region for inactivation as it contains elements used for promotion and enhancement of transcription. When HIV undergoes reverse transcription, the 3′ U3 region is copied to the 5′ end of the genome. Therefore, by deleting critical portions of the 3′ U3 region in the transfer vector including the TATA box, Sp1, NFκB, and NFAT sites, self-inactivating modifications would be incorporated into both long terminal repeats of the integrated vector. These deletions would render the long terminal repeats incapable of transcription.

Third-generation lentiviral vectors were then produced which did not require the HIV Tat protein for production [67]. Tat is the transcriptional activator for integrated HIV proviral DNA and is required for HIV replication. However, modifications were made by deleting the 5′ long terminal repeat U3 region and replacing it with a strong polymerase-II promoter, typically from cytomegalovirus which is ubiquitously utilized in human cells. This allowed for Tat-independent transcription of the transfer vector and reduced the number of HIV genes in the vector production system to only three: *gag*, *pol*, and *rev*. With no active LTRs, only three HIV gene sequences, no accessory proteins, and the utilization of a four-plasmid vector production system, great lengths to develop safe and effective HIV-based lentiviral vectors have been made over the recent years.

Other Modifications and Additions

Current HIV-based lentiviral vectors have incorporated other modifications to improve transduction efficiency and transgene expression. One element called the central polypurine tract has been added to lentiviral vectors to increase transduction efficiency of target cells by increasing the levels of nuclear import of the reverse-transcribed vector DNA [68]. A second element, the woodchuck hepatitis virus posttranscriptional regulatory element (WPRE), has been utilized to increase expression of transgenes by increasing levels of the vector transcripts in both the nucleus and the cytoplasm [69]. A third type of modification includes codon optimization. HIV genes are rich in AU sequences which destabilize protein interactions. The stretches of AUs are also not optimal in human DNA and therefore decrease the expression of gag and pol during vector production. By removing AU-rich regions in the *gag* and *pol* genes, these modifications not only stabilize the sequences and increase the expression of gag and pol, but they also remove the requirement for Rev in the vector production, thus eliminating another wild-type HIV gene.

Numerous other delivery systems have been applied for HIV gene therapy including the Sleeping Beauty transposon system [70], DNA transgene expression plasmids, and cell-specific targeting vectors; however, retroviral and lentiviral vectors have mostly been used for HIV stem cell gene therapy, especially in clinical trials. Moloney murine leukemia viral vectors were first utilized in the initial HIV stem cell gene therapy clinical trials but have recently been replaced by HIV-based lentiviral vectors due to safety concerns and increased stem cell transduction efficiencies, as described above.

Anti-HIV Genes

As mentioned previously, numerous anti-HIV genes have been developed over the past decades, ever since the first concept of gene therapy to treat HIV infection was envisioned. Genes designed to inhibit HIV infection were aimed at either a viral function or a cellular process required for HIV infection or replication and have reached greater than 4 logs of viral inhibition as measured in vitro HIV challenge assays. The HIV life cycle consists of many steps that can be utilized by anti-HIV genes for effective HIV replication disruption including attachment and fusion, entry, uncoating, reverse transcription, nuclear import, integration, proviral transcription, assembly, and budding.

Protein-based strategies in the development of anti-HIV genes include trans-dominant negative mutants, intrakines, single-chain antibodies, zinc-finger nucleases, and retroviral restriction factors. Transdominant proteins are designed to mimic a portion of the function of a normal HIV protein but lack other critical functions necessary for HIV replication. One example of an anti-HIV transdominant negative protein is called RevM10 [6]. This protein retains two of the wild-type Rev functions in binding the RRE and can form multimers; however, it lacks export capabilities. Other examples include Gag mutants and Tat/Rev fusion proteins.

During normal viral attachment and fusion to target cells, HIV utilizes a major receptor, CD4, and a coreceptor, two of the major ones being CCR5 and CXCR4. These receptors are all surface receptors on the HIV target cells. If ligands to these surface receptors were expressed exclusively inside the cells, these ligands would bind to the HIV cellular receptors prior to their expression on the cell surface and sequester them, thus decreasing available receptors on the membrane of cells for HIV to attach to [71].

Single-chain antibodies can be produced which are specific either for an HIV protein or for one of the critical cell surface proteins required for HIV attachment. By expressing a single-chain antibody specific for an essential protein, a block of function or processing will occur. Examples of previously used single-chain antibodies include those targeted to CCR5, Rev, Gag, and integrase [72].

Zinc-finger nucleases are artificial enzymes which are made by fusing a DNA-binding domain to a DNA cleavage domain. These proteins can be targeted to any DNA sequence, and upon binding, cleavage of the DNA occurs. After repair of the damaged DNA, there is usually a mistake made from the original DNA sequence. Zinc fingers have been engineered to target the cellular coreceptor CCR5 as well as HIV genes [73].

A newer class of protein strategies for anti-HIV genes includes the retroviral restriction factor TRIM5α and APOBEC3. These mechanisms are innate forms of retroviral restriction and have recently been applied to prevent HIV infection using gene therapy techniques. TRIM5α of old-world monkeys including African green monkeys and rhesus macaques are capable of blocking HIV infection due to specific amino acid sequences located in their protein sequence. Human TRIM5α is capable of blocking infection of other retroviruses including simian immunodeficiency virus, feline immunodeficiency virus, and other murine retroviruses, however not HIV. Research has identified the critical amino acids from old-world monkey TRIM5α isoforms required for HIV restriction. By mutating the human isoform of TRIM5α through incorporation of the critical HIV-restricting amino acids from the rhesus macaque isoform, a chimeric version of TRIM5α was generated. This chimeric isoform is capable of inhibiting HIV infection prior to integration, and therefore its use as an HIV gene therapeutic is highly advantageous [59].

Members of the APOBEC family, particularly APOBEC3G and 3F, are known to be HIV restriction factors. APOBEC3G is a cytidine deaminase and is packaged into newly formed HIV virions. During reverse transcription, it deaminates minus strand viral cDNA prior to integration and destroys sequence integrity [74].

RNA Strategies

Besides protein-based anti-HIV strategies, RNA molecules have also been a major source for highly potent anti-HIV genes. These include antisense RNAs, ribozymes which are RNA enzymes, RNA aptamers and decoys, and small interfering RNAs [75]. Antisense molecules were one of the first RNA molecules to be used as anti-HIV genes by expressing the complementary strand of an HIV transcript including tat, rev, or integrase. RNA decoys act as RNA homologues to known HIV RNA secondary structures including the TAR region and the RRE [54]. When expressing a TAR decoy, the HIV viral Tat protein which is critical for HIV replication is sequestered away from its normal function and proviral transcription ceases. Ribozymes are another class of RNA molecules which can be used as anti-HIV genes [76]. These RNA molecules are self-cleaving, and when engineered to recognize a complementary viral or cellular transcript, the ribozymes will cleave the transcript so no translation of protein will occur. Ribozymes have been designed to target the cellular coreceptor CCR5 and the HIV transcripts Tat, Rev, and Gag.

The most recent RNA strategy utilizes the innate antiviral defense of RNA interference to design small interfering anti-HIV RNAs. This process of sequence-specific posttranscriptional gene silencing allows for the design of 18–26 base pair long small RNA molecules which are complementary to any desired transcript. Upon binding of the antisense strand of the small interfering RNA, the targeted transcript is cleaved and protein translation is blocked. Numerous anti-HIV small interfering RNAs have been developed including those targeting the cellular coreceptor CCR5 and all of the HIV transcripts including Tat and Rev. As this mechanism is very sequence specific, efforts have been made to target highly conserved regions of HIV's genome for targeted degradation [77].

A number of the abovementioned anti-HIV genes have proceeded to human clinical trials. Among the first genes to be tested in HIV-infected patients were single RNA-based anti-HIV genes, a TR/Tat ribozyme and an RRE decoy, and the protein-based RevM10 gene [37, 39]. Other HIV gene therapy trials using single genes followed including a tat-vpr ribozyme and an antisense RNA targeted to the HIV envelope gene [78, 79]. These trials, however, utilized single anti-HIV genes. As HIV is highly mutatable, a combination of anti-HIV genes is more likely to inhibit HIV infection at a greater level and to also decrease the chance of forming escape mutants. A recent human clinical trial and the first HIV stem cell gene therapy clinical trial utilizing a lentiviral vector was conducted with a triple combination of anti-HIV genes which included a CCR5 ribozyme, a tat/rev short hairpin RNA-interfering molecule, and a TAR decoy [60]. The clinical trials will be described in more detail in a later chapter.

Vector Production

Lentiviral vectors are produced mostly in human embryonic kidney (HEK) 293T cells through transfection with the abovementioned vector production plasmids. These cells are ideal for lentiviral vector production since they are a transformed cell line allowing them to grow continuously, and they are also easily transfectable with multiple plasmids. Transfection of HEK-293T cells can be achieved by calcium phosphate transfection or by lipofection. Vector titers vary greatly in orders of magnitude depending on the health of the packaging cells, the generation of lentiviral vector used, and the complexity of the transfer vector requiring packaging. Three plasmid systems will afford higher titers due to the ease of transfecting three plasmids compared to four plasmids. Also, by adding the central polypurine tract to the transfer vector, higher titers and transduction efficiencies can be achieved.

Various methods of vector concentration have also been applied, both to increase stability of the vector and to achieve higher concentrations. Two widely used methods are ultracentrifugation and ultrafiltration. With the incorporation of VSV-G as a pseudotyping envelope, the stability of the lentiviral vector particles increased and can be centrifuged at high speeds ($3 \times 10^5 \times g$ for 3 h at 4 °C). This process increases the concentration of vector particles at least 100-fold. Even though stability of the vector particles increased by pseudotyping with VSV-G, high-speed centrifugation can lower achievable vector titers due to destruction of vector particles. Ultrafiltration is a second method of concentrating lentiviral vectors which is less strenuous and less time consuming. Lentiviral vectors are filtered in vessels containing micropores applying low-speed centrifugation ($400 \times g$ for 30 min) which retains most, if not all, intact vector particles produced in the packaging cell line while removing debris.

The next stage in vector manufacturing is tittering of the concentrated vector prior to transduction of target cells. This allows for quantitative transduction efficiencies at the desired multiplicity of infection. Lentiviral vector tittering is typically performed on HEK-293T cells through serial dilutions and quantifying the numbers of transduced cells. This titer provides a transducing titer, allowing for calculations of multiplicity of infection. Vectors for preclinical and proof-of-concept research usually incorporate a reporter gene such as enhanced green fluorescent protein or neomycin resistance to track transduced cells by flow cytometry or immunostaining. Other methods to quantify vector titers are p24 antigen enzyme-linked immunosorbent assays (ELISAs) and quantitative real-time polymerase chain reactions with DNA oligonucleotide primers specific for a vector sequence either the therapeutic gene or the vectors psi signal. A major problem with p24 ELISAs, however, is not being able to distinguish transducing from defective vector particles which are not capable of transducing cells.

In summary, many laboratories have contributed over the last decades to building and improving the safety and effectiveness of HIV gene therapy vector systems. With modifications ranging from multiple plasmid transfection systems to highly efficient vector packaging/concentration systems and new highly potent anti-HIV genes, HIV-based lentiviral vectors have now become one of the most important delivery systems for HIV gene therapy and, particularly, HIV stem cell gene therapy.

Chapter 5
Stem Cells for HIV Gene Therapy

HIV is a chronic infection and is mostly characterized by a prolonged disease progression where the virus infects human CD4+ cells and causes immune system destruction and dysregulation, both directly through cell killing and indirectly with bystander cells. Therefore, when developing a durable anti-HIV gene therapy approach, it would not be practical to focus on treating patients with anti-HIV gene-expressing mature immune cells, such as T cells. This type of therapy would only have a transient effect as the gene-modified T cells have a defined half-life and will eventually die off. Viral reservoirs would then be able to re-infect newly made HIV-susceptible immune cells and patients would continue to be infected. Multiple anti-HIV gene-modified T cell infusions over the life of the patient may circumvent these problems; however, this would be a costly and troublesome protocol.

HIV only infects CD4+ cells of the immune system, and all of these cells come from a single progenitor called the hematopoietic stem cell [80]. Every person, young and old, has these stem cells, and they are the reason why we are capable of constantly producing new immune system cells for the rest of our lives. Therefore, by utilizing hematopoietic stem cells (HSCs) which are capable of making all blood cells in the body for HIV gene therapy, a prolonged and possibly one-time treatment with anti-HIV gene-expressing cells could be achieved. HIV does not infect true self-renewing and self-repopulating HSCs which allow them to be utilized as uninfected cells for treatment for HIV [80].

Hematopoietic stem cells are multipotent stem cells and reside in the bone marrow, the majority of which exist in the G0 phase. Once they enter their cell cycle, they can undergo either asymmetric or symmetric division giving rise to daughter HSCs or blood cell progenitors. These blood cell progenitors can then further differentiate into HIV-susceptible cells, including CD4+ T cells, monocytes, macrophages, dendritic cells, and brain microglia. HSCs which are characterized by a CD34+/CD38− phenotype are both self-renewing and self-repopulating which means that they will not only continuously make the abovementioned immune cells for the life of an individual, but they will also make more of themselves.

G. Bauer and J.S. Anderson, *Gene Therapy for HIV: From Inception to a Possible Cure*, SpringerBriefs in Biochemistry and Molecular Biology, DOI 10.1007/978-1-4939-0434-1_5, © Gerhard Bauer and Joseph S. Anderson 2014

If engineered to become HIV-resistant with various methods including vector transduction, zinc-finger modification, or episomal DNA gene expression, a prolonged and self-renewing immune system capable of inhibiting HIV infection could be generated. This strategy, therefore, has great potential to control HIV infection in patients and may possibly one day eradicate the virus from HIV-infected patients.

In 2007, a groundbreaking procedure was performed on an HIV-infected patient, now referred to as the "Berlin patient," who had developed acute myeloid leukemia. To cure the patient's leukemia, he had to undergo an allogeneic bone marrow transplantation after a fully ablative chemotherapy. The bone marrow transplant, however, did not use just any hematopoietic stem cells. The stem cells used were from an HLA-matched donor and were naturally resistant to HIV infection due to a homozygous mutation in the CCR5 gene called delta-32 which prohibits cell surface expression of CCR5. CCR5 is a critical coreceptor utilized by macrophage tropic strains of HIV to attach and fuse with HIV-susceptible cells. Only around 1 % of people of the northern European descent are homozygous for this allele which had initially aided in the survival of the Black Plague. To this date, the "Berlin patient" has been the only documented case of an HIV-infected patient being cured of HIV [81]. This was accomplished by utilizing HIV-resistant hematopoietic stem cells. This procedure, however expensive and involving a potentially lethal conditioning regimen, did set the stage for the possibility of developing an actual cure for HIV.

One major problem with trying to apply this procedure for everyone infected with HIV worldwide is that it would be extremely difficult to find HLA-matched donors who are homozygous for the CCR5 delta-32 allele. Also, using only CCR5 negative cells would not inhibit other strains of HIV which use other coreceptors for attachment and fusion, such as CXCR4-tropic strains. HIV is also highly mutatable due to its error-prone reverse transcriptase and, therefore, may mutate around the CCR5 delta-32 block if allowed to reverse transcribe and replicate.

HIV stem cell gene therapy has a great potential to mimic the outcome observed with the "Berlin patient." When utilizing HIV stem cell gene therapy, patients' own autologous HSCs can be engineered with anti-HIV genes to become HIV resistant. CD34+ HSCs reside in the bone marrow and can be mobilized into the bloodstream of patients after administration of granulocyte colony-stimulating factor (G-CSF). Once in the blood, large quantities of HSCs can be collected safely by apheresis which avoids the painful procedure of bone marrow harvests. This process has been shown to be safe and has become a routine method applied for bone marrow transplantation. Once the product is collected by apheresis, CD34+ HSCs can be isolated by immunomagnetic bead separation and are easily readministered via intravenous injection into patients for engraftment and long-term multi-lineage hematopoiesis.

For HIV stem cell gene therapy, once the CD34+ HSCs are purified from the apheresed product, they can be genetically modified to render them HIV resistant. Upon purification, HSCs are cultured ex vivo with a cytokine cocktail containing stem cell factor (SCF), thrombopoietin (TPO), and flt3-ligand to expand the population. This culture method allows the cells to be stimulated which enables better transduction efficiency when utilizing viral vectors for genetic modification. For ex vivo expansion to be relevant in HIV stem cell gene therapy protocols, however, the

HSCs would need to maintain their self-renewal and repopulating/engraftment ability. There is still limited understanding of the physiological mechanisms which regulate self-renewal of HSCs. Therefore, anti-HIV gene modification needs to be performed relatively quickly, so as not to leave HSCs in culture too long and to compromise their self-renewal abilities. Anti-HIV genes can then be transduced into the stem cells which thereafter can either be frozen and stored for future transplantation or directly reinfused into patients.

For HIV-infected patients who have developed AIDS-related cancers including B cell lymphomas and leukemias, a full bone marrow ablation is required using chemotherapy in order to cure their cancer. HIV stem cell gene therapy can then be "piggy backed" on this procedure by inserting the anti-HIV genes into the purified CD34+ HSCs which were apheresed from the patient prior to receiving chemotherapy for marrow ablation. HIV-resistant stem cells can be reinfused into patients to reconstitute their hematopoietic system. Their cancer will not only be cured, but patients will also generate an HIV-resistant immune system. For HIV infected patients who have not developed cancer, a low-dose chemotherapy ablation regimen can be used. As observed with ADA-severe combined immunodeficiency patients in recent clinical trials, only a reduced intensity conditioning regimen needs to be applied for engraftment of gene-modified cells to occur [16]. Therefore, HIV-infected patients can receive a nonlethal dose of chemotherapy to make space in the bone marrow for the anti-HIV gene-modified cells. This procedure may need to be performed multiple times, however, to engraft enough HIV-resistant stem cells, but it would be a less toxic procedure than full bone marrow ablation.

Genetic modification of CD34+ HSCs for HIV stem cell gene therapy was initially performed with retroviral vectors (Moloney murine leukemia virus). Recently, lentiviral vectors have been used to insert anti-HIV genes into the stem cells. Viral vector transductions are performed ex vivo and there have been wide ranges of efficiency levels. Low transduction efficiencies have also led to low in vivo gene marking levels when the anti-HIV gene-modified cells were reinfused back into patients [60]. This has been the biggest problem in achieving any clinical benefit with previous HIV stem cell gene therapy clinical trials. New methods are being developed to increase the levels of anti-HIV gene-expressing cells once transplanted into patients. These include the incorporation of a gene which enables the transduced stem cells to be resistant to an in vivo administered toxin. Methyl guanine methyl transferase (MGMT) has been used as a selection molecule to protect cells from BCNU, benzyl guanine, or temozolomide [82]. Another method utilizes type II inosine monophosphate dehydrogenase (IMPDH2) and mycophenolate mofetil (MMF) [83]. MMF is an immunosuppressive agent as it inhibits IMPHD which is a critical enzyme required for purine nucleoside production. IMPDH2 is not affected by MMF and, therefore, if incorporated in an anti-HIV vector, transduced cells would survive and be selected for upon administration of MMF. These methods would allow for the in vivo selection of anti-HIV gene-transduced cells which would increase the percentages of gene marking and could potentially lead to a clinical benefit in treated patients.

A new technique utilizing zinc-finger nucleases to disrupt cellular genes critical for HIV replication (such as CCR5) has been applied in CD34+ HSCs; however, this method has a low efficiency in gene disruption which would more than likely not have any clinical benefit [73]. Recent work in developing CD34+ HSC-specific targeting vectors has also shown some promising results. Research has involved incorporating stem cell targeting ligands including stem cell factor and thrombopoietin into the envelope of viral vectors [84]. This has allowed the cell-specific targeted transduction of CD34+ HSCs and could potentially be used as an in vivo HIV stem cell gene therapy procedure instead of current transduction protocols which occur ex vivo.

Advantages of HIV gene therapy utilizing HSCs include the possibility of a one-time treatment, controlled and/or constitutive anti-HIV gene expression, and prolonged HIV inhibition after HSC transplantation. Several HIV stem cell gene therapy clinical trials have been performed on HIV-infected patients with viral vectors. These trials will be discussed in a later chapter.

Pluripotent Stem Cells

For HIV stem cell gene therapy to succeed long term and for anti-HIV gene-expressing immune cells to be available for the life of treated patients, hematopoietic stem cells are ideal cell targets to render HIV resistant. HSCs can be obtained from HLA-matched cord blood, allogeneically from an HLA-matched donor, or autologously from a patient's own bone marrow or mobilized peripheral blood. Another source of HSCs, which has recently been exploited for both cellular therapies and gene therapies, is pluripotent stem cells. Unlike multipotency where a progenitor cell has already committed itself to a certain lineage of cells, pluripotency refers to a cell's ability to differentiate into any cell type of the body. Pluripotent cells are also capable of virtually unlimited proliferation, which is a very much wanted characteristic if these cells are to be used as a therapeutic. Pluripotent stem cells include both human embryonic stem cells (hESCs) and induced pluripotent stem cells (iPSCs). These two groups of cells are capable of being directly differentiated into the three germ layers, the endoderm, ectoderm, or the mesoderm, depending under which conditions and in which environment the cells are growing. They can also be a potential source of both autologous and allogeneic cells for experimental and therapeutic use.

Human embryonic stem cells were the first pluripotent cells to be utilized for stem cell and gene therapy approaches for HIV. hESCs are derived from the inner cell mass of the blastocyst which is an early stage embryo in the development of humans. The blastocysts consist of around 50–150 cells and are removed in the process of deriving hESCs. The embryo, however, needs to be destroyed in order to obtain the inner cell mass to derive hESCs and, therefore, has provided for the most controversy in utilizing hESCs for any therapeutic purpose. There have been numerous studies which demonstrated that hESCs can be differentiated towards the hematopoietic lineage and that this process mimics embryonic hematopoiesis [85]. This not only allows hESCs to be used as a tool to study hematopoiesis but also highlights their potential to be used in stem cell therapies, including for HIV.

Another pluripotent stem cell type which has been employed in stem cell biology is induced pluripotent stem cells (iPSCs). Pluripotency can be transferred to differentiated somatic cells by the addition of specific genetic factors [86]. Since iPSCs can be generated by reprogramming adult somatic cells, this would circumvent the ethical issues in using hESCs. Numerous somatic cells including fibroblasts, T cells, HSCs, and many others have been used as "starter cells" to generate iPSCs. When deriving iPSCs for the directed differentiation of HSCs, numerous reports have demonstrated that using HSCs as the "starter cell" results in a higher efficiency of functional HSCs generated due to epigenetic programming [87]. The first studies to generate iPSCs utilized the expression of Oct4, Sox2, Klf4, and c-Myc to induce a pluripotent state in human fibroblasts. However, recent studies in generating iPSCs have involved the use of combinations of both chemical inducers of pluripotency and one or two of the reprogramming genes. Other new methodologies include the transfection of plasmid DNA, reprogramming mRNAs, and microRNAs. Normally, a fully differentiated end-stage cell is not able to be differentiated into other cell types. However, upon reprogramming into pluripotency, mature cells can be dedifferentiated to resemble hESCs. iPSCs are almost identical to hESCs in their characteristics of being pluripotent, the ability to directly differentiate them into specific cell types, their virtually unlimited proliferation potential, their cellular phenotype in culture, and their ability to form teratomas when injected into immunodeficient mice. Various cell types have been derived from iPSCs including hepatocytes, mesenchymal stem cells, neurons, and hematopoietic progenitor cells (HPCs).

In vitro differentiation of hESCs and iPSCs into HPCs has been demonstrated using a variety of methods both with stromal cells and embryoid body formation. Stromal cell layers which have been utilized for the differentiation of pluripotent cells into HPCs include mouse S17 and OP9 cells and human mesenchymal stem cells. These stromal cells have been used due to their ability to mimic the embryonic environment of hematopoietic development. Along with culturing pluripotent cells on stromal cells, the cytokine bone morphogenic protein-4 (BMP-4) has been added to the differentiation cultures to increase the levels of HPCs obtained. After 7 days of coculturing, around 20 % of pluripotent cells differentiate into hematopoietic progenitors with a CD34+ phenotype. However, when pluripotent stem cells are cultured in embryoid bodies, the addition of a variety of other cytokines is required for HPC differentiation. These include stem cell factor, IL3, IL6, Flt-3, BMP-4, and vascular endothelial growth factor. The requirement for these additional cytokines is due to the absence of the stromal layers which would normally supply a number of these factors. Using this method, again around 20 % of pluripotent cells differentiated into CD34+ hematopoietic progenitors. Advantages of the embryoid body differentiation method include the removal of stromal layers which can contaminate HPC cultures, the use of a xeno-free system, and defined culture conditions.

To determine the complete functionality of hESC- and iPSC-derived HPCs, the in vivo engraftment and multi-lineage hematopoietic potential needs to be evaluated by transplanting these cells into immunodeficient mice. Numerous reports have described the in vivo engraftment of hESC- or iPSC-derived HSCs by detection of human CD45 3–6 months post-transplantation. Levels of engraftment have ranged from 1 to 16 %. Secondary transplant recipients have also confirmed the successful derivation of true engrafting HSCs. For future therapeutic applications, a significant improvement in engraftment levels needs to be achieved [88]. Also, methods to purify HSCs from pluripotent cultures need to be free from pluripotent cells to avoid oncogenesis. These initial results, however, hold future potential in utilizing pluripotent stem cells for hematologic disorders including HIV [87].

As HIV infects CD4+ cells of the immune system, cell replacement therapies based on reconstituting the hematopoietic compartment with hESC- and iPSC-derived HSCs have considerable promise. Functional immune cells including macrophages, dendritic cells, natural killer cells, and B cells have been derived from hESCs and iPSCs and have been studied in proof-of-concept experiments for immunotherapies. Anti-HIV genes can be inserted into hESCs and iPSCs, and upon directed differentiation towards the hematopoietic lineage, HIV-resistant HSCs can be generated. These HSCs can then be derived into immune cells in vitro or transplanted into immunodeficient mice to evaluate the in vivo engraftment and multilineage hematopoiesis. Previous studies have demonstrated that both phenotypically and functionally normal HIV-resistant macrophages and dendritic cells have been generated from these pluripotent stem cell types.

Current HIV stem cell gene therapy procedures require apheresis of a patient's mobilized peripheral blood stem cells, ex vivo transduction of purified HSCs, and transplantation of the genetically modified cells back into the patient. However, a major drawback with this protocol is that the number of cells obtained for ex vivo manipulation is limited. By introducing anti-HIV genes into hESCs or iPSCs, it is possible to generate a virtually unlimited and large supply of patient-specific anti-HIV HSCs for use as a cellular therapy. Also, this would allow for the generation of a larger quantity of anti-HIV HSCs from cell banks if future transplantations, whether autologous or allogeneic, are required [89].

Anti-HIV gene-expressing pluripotent stem cells have the potential to improve upon current HIV stem cell gene therapy procedures by providing HSCs which all contain anti-HIV activity, thus alleviating the current problems of low transduction efficiencies and low in vivo gene marking. They also hold great potential to bridge the gap between current protocols and the successful suppression of HIV replication in an HIV-infected patient who received an allogeneic bone marrow transplant from a donor homozygous for the CCR5 Δ32-bp deletion. Another advantage with using anti-HIV pluripotent stem cells for HIV gene therapy is that each cell line can be fully characterized for safety including the analysis of integration sites of the anti-HIV genes. For future use of pluripotent stem cells for cellular therapies, integration sites of the anti-HIV vector can be defined, and the cells which have "safe harbor" sites can be selected, expanded, and further characterized for their anti-HIV efficacy.

Chapter 6
Animal Models Used in HIV Gene Therapy

Before anti-HIV gene therapy vectors can be applied in human clinical trials, the in vivo safety and efficacy of both the anti-HIV genes and the therapeutic vector need to be evaluated. As HIV can only infect human cells, animal models of HIV disease need to be developed where the transplantation of human cells is allowed. These cells can include either CD34+ hematopoietic stem cells or primary peripheral blood mononuclear cells. For HIV gene therapy, a number of preclinical animal models have been developed including humanized mouse models and nonhuman primate models.

Murine Models

To evaluate HIV gene therapeutics, novel approaches utilizing immunodeficient mouse models have been applied to preclinical research. As these mice are immunodeficient, they are allowed to accept grafts of human cells and tissues without the worry of rejection. Initially, severe combined immunodeficient (SCID) mice were engineered to harbor human immune cells. The first two mouse models developed were the SCID-hu *thy/liv* mouse model and the SCID-hu-PBL model [90]. SCID-hu *thy/liv* mice are generated by implanting human thymic tissue along with human fetal liver in the kidney capsule of SCID mice. The thymic tissue supplies the architecture for human T cell development while the liver supplies the human hematopoietic stem cells which will develop into phenotypically and functionally normal T cells during lymphopoiesis. These implanted thymic grafts can sustain human T cell development for greater than 1 year. This mouse model allows for the evaluation of HIV stem cell gene therapies upon transplantation of anti-HIV gene containing CD34+ HSCs purified from cord blood, bone marrow, or human fetal liver. Development of T cells in the thymus can be studied along with in vivo challenge assays upon injection of HIV into the thymic grafts. HIV infection leads to human thymocyte depletion and changes the architecture of the thymic grafts resembling

G. Bauer and J.S. Anderson, *Gene Therapy for HIV: From Inception to a Possible Cure*, SpringerBriefs in Biochemistry and Molecular Biology, DOI 10.1007/978-1-4939-0434-1_6, © Gerhard Bauer and Joseph S. Anderson 2014

normal human HIV infection. Limitations of this model, however, include minimal hematopoiesis as only T cell development can be studied and the lack of development of a complete human immune system to study chronic HIV infection. Also, CD34+ hematopoietic stem cells need to be directly injected into the thymic grafts; thus, trafficking and homing of stem cells cannot be studied.

The SCID-hu-PBL mouse model offers both advantages and disadvantages compared to the SCID-hu *thy/liv* model [90]. In this model, mice are transplanted with primary human peripheral blood mononuclear cells, essentially reconstituting the mice with a human peripheral immune system. These mice are capable of accepting the human cells which can later be found in lymphoid organs including the spleen and bone marrow as well in the peripheral blood. Upon HIV infection, humanized SCID-hu-PBL mice display hallmarks of normal HIV disease progression including human CD4+ cell depletion and an increase in plasma viremia. Immune system activation, surveillance, killing, and antibody production against HIV and HIV-infected cells can be studied, albeit for a short time. However, as these mice are injected with mature immune cells, stem cell gene therapy is difficult to evaluate using the SCID-hu-PBL model. Also, this mouse model, although immunodeficient, can spontaneously develop mouse T and B lymphocytes and mouse natural killer cells as well. This prevents the establishment of prolonged human immune cell engraftment, and therefore, characteristics of chronic HIV infection cannot be studied.

The next generation of humanized mouse models led to the development of bone marrow–liver–thymus (BLT) mice which are made by transplanting human lymphoid tissue into nonobese diabetic (NOD)/SCID mice. NOD/SCID mice are used to generate the BLT mice due to their lower levels of murine natural killer cells. The lower numbers of NK cells allow for the increased engraftment of human CD34+ hematopoietic stem cells. First, a sandwich of 1–2 mm pieces of human fetal thymus and liver tissue is transplanted in the kidney capsule of NOD/SCID mice. Next, mice are sublethally irradiated with gamma irradiation which facilitates engraftment of the transplanted human CD34+ cells. In the BLT mice, the transplanted fetal thymus/liver tissue develops into a functional human thymus and allows for T cell lymphopoiesis in the context of human thymic stroma and not on murine thymic epithelium. This allows for the development of T cells which are HLA-restricted and function as normal human T cells [91].

Both human innate and adaptive immune system cells can be found in engrafted BLT mice which display the presence of CD4+ and CD8+ T cells, B cells, monocytes, dendritic cells, NK cells, and gamma delta T cells in the peripheral blood as early as 4–8 weeks posttransplant. The relative levels of each of these subsets of immune system cells found in the BLT mice are very similar to the ratios found in human blood. Also, just as found with human T cells, those developed in the BLT mice have a wide array of Vβ-T cell receptors. Lymphoid organs of human CD34+ cell engrafted mice also display a normal distribution of human immune cells. T cells, B cells, monocytes/macrophages, and dendritic cells populations have been found in normal ratios in the spleen, bone marrow, lungs, liver, and gastrointestinal tract demonstrating multi-lineage human hematopoiesis. Human immune cells including macrophages, B cells, and T cells have also been identified in the vaginal mucosa of engrafted BLT mice. These were important findings as these mice could be potentially utilized to study vaginal transmission of HIV, as will be discussed below.

The most recent humanized mouse models developed for the study of HIV include the NOD/SCIDγc$^{-/-}$ and the RAG1 or 2$^{-/-}$γ$^{-/-}$ mice [92]. These two mouse models share a similar phenotypic mutation where the common gamma chain of the IL2 receptor is knocked out abrogating normal T and B cell development and activity. This mutation is a component of the IL2, IL4, IL7, IL9, IL15, and IL21 receptors which renders the mouse's immune system severely handicapped. The RAG1 and RAG2$^{-/-}$γ$^{-/-}$ mice also are deficient in their recombinase activating genes (RAG). These mutations do not allow any leakiness of T or B cell development. NK cell activity is also prevented due to the lack of IL15 signaling from the gamma chain mutation. Due to the mutations made to the mouse genome in generating these immunodeficient mice, the best engraftment of human CD34+ hematopoietic stem cells has been observed with these strains. To generate humanized mice using the NOD/SCIDγc$^{-/-}$ or the RAG$^{-/-}$γ$^{-/-}$ mice, sublethally irradiated (200–300 rads, respectively) newborn pups (2–5 days old) are transplanted intrahepatically with at least 10^5 human CD34+ cells. During the first week after the birth of the mice, hematopoiesis occurs from the liver, and therefore, upon injection of human CD34+ cells, homing and engraftment of the various lymphoid organs can occur with the human stem cells. Multi-lineage human hematopoiesis of all the major immune cells occurs in engrafted mice in the peripheral blood, spleen, thymus, bone marrow, lymph nodes, liver, lungs, gastrointestinal tract, and in the vaginal/anal mucosa. Transplanting newborn pups with human CD34+ cells allows for a higher level of human cell engraftment then if the mice are transplanted as adults. This is due to the adult murine immune cells already engrafting and establishing a niche in the lymphoid organs. Also, this method of transplanting human CD34+ cells into newborn pups is a relatively simple procedure compared to the generation of BLT mice and has made these two models widely accepted in many laboratories. A major difference, however, is that in the NOD/SCIDγc$^{-/-}$ and the RAG$^{-/-}$γ$^{-/-}$ mice, T cell development occurs on mouse thymic stroma where human thymic stroma is involved in humanizing BLT mice.

These humanized mice are novel in vivo HIV stem cell gene therapy tools which allow laboratories to evaluate both the safety and efficacy of their gene therapy strategy [93]. Safety of a genetically engineered stem cell will always be the primary concern when moving a therapeutic candidate towards human clinical trials. There is always the possibility that the introduction of anti-HIV genes or alterations to the host genome will cause aberrant effects which may affect homing and engraftment of human CD34+ stem cells, multi-lineage hematopoiesis of the stem cells, engraftment in lymphoid organs, or even the functionality of the developed immune cells. Therefore, once genetically altered, either through vector transduction, zinc finger modification, or episomal DNA expression, stem cells can be transplanted into these mouse models to evaluate safety through engraftment, multi-lineage hematopoiesis, and functionality.

T cell development can be studied in the SCID-hu *thy/liv*, BLT, NOD/SCIDγc$^{-/-}$, or the RAG$^{-/-}$γ$^{-/-}$ mice; however, only studies using the SCID-hu *thy/liv* or BLT mouse models can be performed in the context of human thymic stroma. The other two models utilize murine thymic stroma. However, as compared to the SCID-hu *thy/liv* model, all of the other humanized mouse models enable evaluation of complete human immune system multi-lineage hematopoiesis with reconstitution levels tested in the peripheral blood and lymphoid organs. The newer generations of humanized mice also allow for the evaluation of the functionality of all HIV gene-modified immune cells and do not limit studies to strictly T cells [93]. Therefore, when evaluating a new HIV gene therapeutic candidate, safety studies including phenotypic and functional analyses can be performed utilizing these humanized mouse models.

HIV Infections in Humanized Mice

HIV is restricted to infecting human CD4+ cells and cannot infect murine immune cells. However, upon humanization, the abovementioned mouse models are then capable of supporting HIV replication and display characteristics hallmark of normal HIV infection including human CD4+ cell depletion and an increase in plasma viremia. These humanized mice support infection with a wide array of HIV strains including CCR5-tropic, CXCR4-tropic, dual-tropic, and even primary isolates of HIV-1. This is an important characteristic of these in vivo HIV disease models as they allow for the study of HIV evolution and changes in both the viral sequence and its adaptation to the gene therapeutic candidate being studied.

Differences in HIV infection models between humanized mice, depending on what area of HIV research is being studied, will determine which model a laboratory will want to use. For example, with the SCID-hu *thy/liv* model, a direct thymic injection of HIV needs to be performed to infect the T cells. This, however, is not physiologically normal and is maintained in a semi-closed system, and there are no other immune cells available for study or to mount an immune response against HIV-infected cells. The SCID-hu *thy/liv* model does harness some advantages including the use of human thymus architecture which allows for HLA restriction and study of the pathological consequences of a human thymus during HIV infection. It also allows for the study of immature and naive CD4+/CD8+ double positive T cells in the course of an HIV infection.

The new generation of humanized mice allows for multiple routes of infection with HIV which are more physiologically normal. As compared to the SCID-hu *thy/liv* mouse model which only enables a direct injection of HIV into the thymic graft, the BLT, NOD/SCIDγc$^{-/-}$, or the RAG$^{-/-}\gamma^{-/-}$ mice can be infected with HIV either intravenously, intraperitoneally, or at mucosal sites of infection including the vaginal and anal mucosa.

These mice also allow for a more in-depth study of HIV infection as multi-lineage human hematopoiesis occurs in these models. The sites of HIV replication in these mouse models are also similar to that seen in human HIV infection. HIV replication occurs mostly in the spleen, lymph nodes, and in the thymus. The other lymphoid organs including the bone marrow, gastrointestinal tract, vaginal/anal mucosa, and the lungs harbor sites of HIV replication, albeit at much lower levels.

Upon infection of the humanized mice with HIV, human CD4+ cell depletion can be observed within the first couple of weeks depending on the dose of HIV injected and is usually analyzed by flow cytometry [61]. A decline of human CD4+ cells is observed in every humanized mouse model with mostly all HIV strains and occurs no matter what route of infection is used. This loss of CD4+ cells occurs in the peripheral blood and also in the lymphoid organs including the spleen, lymph nodes, and thymus [94]. CD4+ cell rebound has been reported in a number of cases; however, this has coincided with a drop in HIV plasma viremia which may allow for the human CD4+ cells to rebound. Viremia is usually measured by quantitative real-time PCR since p24 antigen ELISAs are not always sensitive enough to detect virus. The peak levels of viremia generated in infected humanized mice are around 10^7 copies per milliliter and rise to that level within the first 6 weeks postinfection. These levels of viremia are maintained throughout the course of infection.

Both innate and adaptive immunity can be studied in these humanized mice as they are able to generate the major components of a human immune system. Human HIV-specific CD4+ and CD8+ cytotoxic T cells have been found in HIV-infected humanized mice as measured by ELISPOT for interferon gamma. After prolonged infection with HIV, memory T cells have also been found. IgM and IgG antibody productions by human B cells specific to HIV gp120 and Gag have been detected in these mice. Also, both neutralizing and non-neutralizing antibodies can be generated following HIV infection.

These humanized mouse models open up an array of both safety and efficacy studies to evaluate new HIV gene therapeutic candidates. CCR5 inhibitors including small-interfering RNA, ribozymes, zinc fingers, and single-chain antibodies can be evaluated to determine if the knockdown of CCR5 expression alters immune cells homing, engraftment, or functionality. Also, the expression of foreign anti-HIV molecules like a modified human TRIM5α, a TAR decoy, or a RevM10 protein alters the physiology of vector-transduced CD34+ stem cells or their immune cell progeny. By using parametric multicolor flow cytometry, the phenotypic profile of anti-HIV gene modified can be studied in both the peripheral blood cells and in the engrafted lymphoid organs. Long-term engraftment of gene-modified stem cells can also be evaluated. Upon engraftment in the bone marrow of transplanted mice, these human stem cells should be capable of maintaining human hematopoiesis for the life of the mouse. Thus, long-term safety studies can be performed on anti-HIV genetically modified stem cells to detect aberrations in sustained engraftment or in the development of populations of immune cells over time. Tumorigenicity studies can also be performed on anti-HIV gene-modified cells to determine whether the genomic alteration (whether vector insertion or anti-HIV gene expression) had induced oncogenesis [95]. Anti-HIV gene-modified cells can be sorted ex vivo and analyzed for their functionality including HIV-specific CD4+ or CD8+ T cell responses. Microarray analyses can also be performed on various immune cell subsets to determine the total genomic expression profiles of anti-HIV gene containing cells.

As far as efficacy testing, humanized mice transplanted with anti-HIV gene-modified stem cells can be evaluated for their ability to maintain normal levels of human CD4+ cells, reduce plasma viremia, and display a selective survival advantage of the HIV-resistant immune cells [93]. If progressed to human clinical trials, an anti-HIV gene therapy candidate must have the ability to block HIV infection in the face of a viral load. The hallmark of AIDS is a decline of human CD4+ cell levels below 200 cells/μl. Therefore, the maintenance of normal CD4+ cell levels during HIV infection is a key ability for any new HIV gene therapy strategy. Detectable viremia may or may not be a critical determinant for human clinical trial success.

The reason why HIV-infected patients progress to AIDS is due to a low CD4+ cell level which opens up the body to opportunistic infections which are normally controlled with a working immune system. Therefore, by maintaining normal CD4+ cell levels, opportunistic infections can be controlled. It is known that long-term nonprogressors with HIV can have detectable viremia. However, due to their genetic makeup, they are able to control HIV infection and maintain normal CD4+ cell levels which allow them to not progress to AIDS even though viral replication is still occurring in their body [96]. Also, in certain strains of nonhuman primates including sooty mangabeys which are naturally infected with the simian form of HIV, SIV, they are able to live with virus replication without developing an AIDS-like syndrome. This is more than likely due to the sooty mangabey's genetic makeup which allows them to maintain normal CD4+ cell levels even while SIV is infecting cells and replicating [97]. Therefore, CD4+ cell levels may be the critical factor in evaluating an anti-HIV gene during HIV infection, and plasma viremia may not be a definitive result of whether the therapeutic candidate will be successful or not. If the maintenance of CD4+ cell levels is critical during HIV infection, then the selective survival advantage of anti-HIV gene-modified cells would also be a critical component to evaluate. As HIV-susceptible cells are being infected and killed off, the anti-HIV gene-modified cells due to their HIV-resistant phenotype should be able to survive, proliferate, expand in the population of immune cells, and take over the immune system. This population of HIV-resistant immune cells would be the contributing factor to the maintenance of normal CD4+ cell levels above 200 cells/μl and stop the progression towards AIDS.

Nonhuman Primates

During the 1990s, several laboratories developed nonhuman primate models of HIV infection. Rhesus macaques would develop AIDS-like symptoms after being infected with different forms of SIV and displayed a similar but not exact pathology or disease progression to humans infected with HIV [98]. Nonhuman primates especially rhesus macaques can play a major role in evaluating the preclinical safety and efficacy of new anti-HIV gene therapy candidates. Nonhuman primates are most like humans in a variety of ways including genetically, immunologically, physiologically, and also in their hematopoietic system. This has been shown by the use of many antibodies which cross-react with nonhuman primate tissues.

However, there are many considerations to be weighed when using nonhuman primate models for HIV infection including ethical issues, the availability of primates, and the high cost of maintaining a colony. Other considerations to take into account when utilizing nonhuman primate models to test anti-HIV gene therapy candidates include genetic factors such as the expression of TRIM5α in old-world monkeys which can potently inhibit transduction with HIV-based lentiviral vectors [99]. SIV-based vectors can be used for studies; however, laboratories would not be testing the actual vector to be used in human clinical trials. Also, SIV vectors do not easily transduce human CD34+ hematopoietic stem cells. Hybrid lentiviral vectors with components from both HIV and SIV have been developed, such as utilizing the SIV capsid with an HIV-based vector. These vectors have been able to transduce nonhuman primate CD34+ stem cells albeit at lower levels than with human CD34+ stem cells. Also, viral challenges would need to be performed with HIV/SIV hybrid viruses termed SHIVs as rhesus macaques are not infectable with HIV. Therefore, resistance to both natural HIV infection and anti-HIV gene functionality could not be specifically assessed. These include both the pathology and viral replication which, in vivo, do not replicate infection with HIV. Gene transfer in the pigtailed macaque model was found to be efficient and can be utilized for safety studies; however, HIV infection is inefficient. Therefore, efficacy studies with the anti-HIV genes would be limited.

Gene therapy studies have performed in the rhesus macaque model system by a number of groups [100]. In one study, a group analyzed the effectiveness of an SIV-specific ribozyme in vector-transduced rhesus macaque CD34+ cells [101]. Protection from infection was observed; however, as stated above, these studies do not evaluate the true anti-HIV gene therapeutic candidate and also do not provide data on HIV challenge assays.

Improvements and advancements continue to be made to nonhuman primate models for HIV infection but need to reach the level where anti-HIV stem cell gene therapy can exactly mimic that of humans.

Chapter 7
Manufacturing of a GMP Grade Product for HIV Gene Therapy

Good manufacturing practice (GMP) is a national standard for the production of safe and efficacious pharmaceuticals. It describes methods used for and the facilities and controls needed for the production of such pharmaceuticals. GMP has not been internationally standardized and, therefore, many jurisdictions follow their own GMP principles and rules which make their worldwide application rather complicated and the international exchange of GMP manufactured pharmaceuticals difficult. "Current" good manufacturing practice or "cGMP" refers to the constant updating of these production standards to keep current with the state of the art.

In the United States, cGMP regulations are set forth in the Code of Federal Regulations Title 21, Parts 210 and 211 [102, 103]. Current minimum standards are described there, with the latest revision as of April 1, 2012. These regulations apply to the manufacture, processing, packing, or holding of a drug to assure that the drug meets the necessary requirements for safety, identity, strength, quality, and purity.

Recently, these standards have been adapted to include "cellular" drugs, human tissues, and cellular and tissue-based products. These regulations are set forth in 21CFR Part 1271 and are also called "Good Tissue Practice" (GTP) [104]. Such regulations are specifically written to prevent circumstances of transmission of human transmissible diseases and to create proper labeling and identification of such products to prevent cross contamination, as many of these products are derived from different tissue donors.

GMP regulations are absolutely vital to guarantee that a therapeutic that will be administered to a human is free from contaminants, including microbiological contaminants, and will not cause harm. Regrettably, incidences in the past prompted these regulations [105].

To be compliant with GMP regulations, in addition to the application of cGMPs during product manufacturing, a GMP facility is needed. This is a manufacturing laboratory under strict environmental control. Controlled parameters include temperature, humidity, particulate matter in the room air, and room pressurization applying pressure gradients. The United States pharmacopeia (USP) states that "high risk manufacturing" should be carried out in a "Class 10,000" manufacturing

G. Bauer and J.S. Anderson, *Gene Therapy for HIV: From Inception to a Possible Cure*, SpringerBriefs in Biochemistry and Molecular Biology, DOI 10.1007/978-1-4939-0434-1_7, © Gerhard Bauer and Joseph S. Anderson 2014

room, while "low risk manufacturing" can be carried out in a "Class 100,000" manufacturing room [106]. The room classification means that less than 10,000 or 100,000 particles greater than 0.45 $\mu m/ft^3$ of air per minute are present. The outside air contains more than 35 million particles of that size in the same volume. Particles are filtered out using "high-efficiency particulate air" (HEPA) filters in the room air supply. Enough air changes per hour need to be applied to keep the particulate count low under dynamic conditions, that means, when personnel is present and working in the controlled environment. Airborne particles may be viable or nonviable, with dirt and dust particles among the nonviable particles and airborne bacteria and bacterial and fungal spores among the viable particles. Viable particles are particularly dangerous for cultured products since the culture conditions may amplify viable particles, if they get inadvertently inoculated. Therefore, strict aseptic conditions have to be applied, with personnel being fully gowned and sterile technique observed at all times.

GMPs also specify the "quality control" (QC) of all controlled parameters, pertaining to the facility, the manufacturing process, and the final product. All steps in the manufacturing process are regulated by standard operating procedures (SOPs); the facility is controlled by SOPs; the receipt, labeling, storage, and shipping of incoming and outgoing materials are regulated by SOPs; and even the writing of an SOP is regulated by a specific SOP. It is not unusual that a well-organized GMP facility has more than 100 SOPs on file and in use. A quality control unit performs the QC of the facility, manufacturing process, and the final product. A separate "quality assurance" (QA) unit is the final authority responsible for the release of a clinical grade product. This unit needs to make sure that all parameters required for product manufacturing and release have been met and that all documentation is available for the particular product. In essence, the QA unit can also be seen as the "quality control of the quality control."

Before a GMP facility can start operating, all of its functions need to be validated, which includes equipment for manufacturing and storage, such as biosafety cabinets, freezers, refrigerators, and incubators. Although the automated monitoring of facility and equipment is normally carried out, manual monitoring of facility and equipment is also required as monitoring equipment failure can occur.

After this description it becomes clear that a GMP facility is a highly sophisticated, technically complicated laboratory that needs to be controlled at all times. Personnel working in such a facility also need to be controlled at all times, as personnel is the most error prone aspect of a GMP manufacturing process. Personnel need to receive appropriate documented training and need to demonstrate proficiency. At all times, during a manufacturing run, a second person, the QC person, needs to be present to check off step by step the individual manufacturing steps that were performed by the manufacturing technician. This assures a controlled and reproducible product.

In order to proceed to clinical applications of stem cell gene therapy for HIV, in essence, two different GMP grade products need to be manufactured: the vector that transfers the anti-HIV genes and the cellular product that receives the anti-HIV genes. The manufacturing of these two different products creates a major dilemma.

Cellular products have traditionally been GMP manufactured in manufacturing rooms that were positively pressurized towards entry and exit rooms, which prevents airborne contaminants to enter the manufacturing rooms. At all times, some manufacturing room air is pushed out under the manufacturing room entrance and exit doors. Cellular products are cultured in culture vessels, and they only pose minimal aerosolization risk and usually do not contain aerosol transmissible contaminants. Therefore, it is not anticipated that manufacturing room air could contain aerosolizable cellular product or product contaminants.

Gene therapy vector, such as the lentiviral vector, is usually concentrated during manufacturing, a process which can create an aerosolizable product containing high concentrations of vector particles. Although replication incompetent, vector particles that are pseudotyped with VSV-G envelope can infect all mammalian cells and permanently integrate their genetic payload. Current concentration protocols applied in our GMP facility can achieve lentiviral vector titers up to 10e10 transducing particles per ml, a considerably high titer. Aerosolization can occur during centrifugation processes, which are often unavoidable for vector manufacturing. To manufacture such a gene therapy vector in a positively pressurized manufacturing room would pose a potential hazard to the outside rooms, as vector particles, if aerosolized, could be pushed out under the manufacturing room doors. Therefore, vector needs to be manufactured in manufacturing rooms that are designed as pressure sinks, similar to BSL3 laboratories. Negative pressure manufacturing rooms have a tendency to become dirty very quickly, as some air from the outside entry or exit rooms, which is not as clean as the air in manufacturing rooms is constantly flowing into the manufacturing rooms. Additionally, in a traditional GMP facility, positive and negative pressure manufacturing rooms were extremely difficult to build using the same air flow system.

This problem has recently been solved in the new GMP facility at the UC Davis Medical Center. It was anticipated that negative pressure manufacturing rooms would be needed for vector manufacturing and positive manufacturing rooms would be required for cellular product manufacturing. A GMP facility was therefore designed that allowed for switchable manufacturing room pressurization, in conjunction with entry and exit areas that served as appropriate air buffer rooms, at the same time enabling one-way personnel, product, and waste flow. Using laminar air flow created by spaced out terminal HEPA filters in the manufacturing room ceiling, large low air returns, and individual air handlers for each manufacturing room in conjunction with a sophisticated air flow and pressure measurement and control system employing remote controlled cone valves, perfectly clean negative pressure areas could be created on demand. Switching of pressurization was validated; exactly prescribed positive or negative room pressure values can be achieved within minutes of the switch which is executed via computer screen buttons.

To manufacture a lentiviral vector suitable for clinical grade gene transductions, the first step is the generation of the HEK-293T producer cell master cell bank. For engineering runs, it is good enough to just use laboratory grade 293T cells. However, the use of these cells for clinical applications is quite restricted. Therefore, certified HEK-293T cells with an appropriate clinical use agreement should be acquired. Master cell bank production occurs by GMP grade expansion of clinical grade 293T cells, using qualified reagents, controlled rate freezing of these expanded cells in a large number of vials, and appropriate QC testing of a representative number of master cell bank vials. Tests must include 14-day sterility, endotoxin, and mycoplasma. Adventitious virus testing is also required.

From this master cell bank, a working cell bank is manufactured and tested and, if passed, used to seed appropriate culture vessels. Culture vessels can be single layer or multilayer tissue culture flasks, cell factories, or other large-scale culture devices. Often, the clinical grade vector is associated with lower titer, since GMP vector production has to follow stringent procedures that emphasize safety (freedom from contamination) but not maximum titer. The great advantage of flask cultures is the immediate visibility of the plated 293T cells. A trained technician can easily determine the quality of the producer cells and seeding density by simple observation. A bad flask can easily be eliminated, if necessary. This helps with optimizing culture conditions and selection of flasks that will yield high titer vector. This is not easily possible in multilayer cell factories. It is much easier to produce high-quality vector in a multiflask system, often with a titer in the 10e10 transducing particles per ml range; however, it is also much more labor intensive when a large quantity of vector is needed. After the producer cells have reached the appropriate density, the 293T cells are transfected with clinical grade plasmids; for a third generation lentiviral vector, this would be the gag/pol plasmid, which provides the viral structural genes, the *env* plasmid which provides the VSV-G gene, and the transfer plasmid, which carries the anti-HIV gene(s). Plasmid manufacturing is often outsourced to specialized companies that can provide high concentrations of pure plasmids, free of contaminants, with particularly low levels of endotoxin. These companies also perform the release testing on these plasmids. After plasmid transfection, which can be as simple as a calcium phosphate precipitation, the media is changed to serum-free media, and vector collection follows after allowing time for proper vector production and secretion from the producer cells. Dependent on the size of the vector preparation, the collected supernatant may be pre-concentrated via tangential flow filtration and treated with benzonase to remove residual plasmid DNA. Final concentration and volume reduction is accomplished by ultracentrifugation. Alternatively, the vector can also be concentrated by spin column filtration. Concentrated vector is stored at −80 °C.

After vector manufacturing, to conform to the FDA requirements, the following tests need to be performed: 14-day sterility, endotoxin, mycoplasma, replication-competent lentivirus, plasmid DNA, benzonase, transducing titer, sequence, and stability (stability testing is done by transducing titer evaluation every 6 months). The limits for these tests are freedom from microbial contamination, endotoxin, and replication-competent lentivirus, undetectable benzonase and plasmid DNA, transducing titer in the 10e9 particles per ml range, sequence as predicted, and stable titer over 6 months or more. A final certificate of analysis will be issued, tabulating the results of all these tests.

To achieve sufficient vector titer, it is often required to perform a number of engineering runs and to optimize the vector production protocol. However, this may sometimes be complicated by an inherent packaging inhibition exerted by the anti-HIV vector. Some of the genetic information of HIV is used to produce the vector particle, and certain anti-HIV genes can interfere with the activity of these genes. Our own anti-HIV vector has therefore been optimized to overcome this problem to consistently yield titers in the high 10e9 per ml range.

The second product that needs to be GMP manufactured for stem cell gene therapy for HIV is the cellular product, autologous transduced hematopoietic stem and progenitor cells. A bone marrow aspirate or mobilized peripheral blood stem cells can serve as starting products. From these products, CD34+ cells need to be isolated. The only clinical grade method of accomplishing this is currently the Miltenyi CliniMacs magnetic bead cell separation system. Ferromagnetic dextran microbeads are coupled to a GMP grade antibody against CD34 and added to the starting product. CD34+ cells will bind to the antibody and through it to the magnetic beads. A very strong magnetic field then allows the cells bound to the beads to be retained in a flow-through column. After the removal of the magnetic field, CD34+ cells then can be eluted. The Miltenyi CliniMacs system is a closed system utilizing clinical grade buffers and reagents [107]. With this system it is possible to isolate CD34+ cells directly from mobilized peripheral blood units without preprocessing, as they are already enriched for white cells. It is not easily possible to isolate CD34+ cells from bone marrow aspirates. These should be enriched for mononuclear cells via Ficoll density gradient centrifugation prior to CD34+ cell isolation. The purity of the CD34+ cell fraction after the magnetic bead separation ranges from 80 to 95 %.

For GMP manufacturing, it is important to utilize built-in control points, which is called "in-process testing." The isolated CD34+ cells are therefore enumerated for CD34+ cells by flow cytometric analysis.

After CD34+ cell isolation, immediate culture of CD34+ cells follows on a RetroNectin (fibronectin fragment CH296) matrix in tissue culture flasks. It is important to provide hematopoietic stem cells with a matrix to attach to during culture in order to facilitate stem cell survival and engraftment [108]. Low levels of the cytokines Flt-3, IL-6, and SCF are used to also help with survival of the stem cell fraction. It is not required to stimulate the stem cell fraction to divide, since lentiviral vector is able to transduce resting cells. All media components used need to be qualified and cytokines and matrix need to be clinical grade. COAs need to be available for all components. GMP grade, tested lentiviral vector is added to the stem cell culture at a multiplicity of infection of approximately 10–20, which is then incubated for only 24 h. After 24 h vector is washed out carefully (three washes are required), samples for release testing are taken, and the cellular product is controlled rate frozen in large and small aliquots and stored in the vapor phase of liquid nitrogen.

The following release tests need to be performed on the cellular product: cell count, viability, sterility, identity and purity (flow cytometric analysis), endotoxin, mycoplasma, transduction efficiency, integrated vector copy number, and potency (e.g., HIV suppression in an in vitro challenge assay). A small aliquot of the frozen product is also thawed to assess thaw viability. The limits of these tests are freedom from microbial contamination and replication-competent lentivirus, cell count within range to provide proper engraftment, viability both after culture and postthaw greater than 70 %, integrated copy number less than 3, transduction efficiency greater than 10 %, CD34+ content similar to the starting product confirming identity and purity, and resistance to HIV challenge in the in vitro challenge assay confirming potency.

The transduction efficiencies achieved with our vector, using this process range from 20 to 50 %.

It has been argued whether the transduced cells should be infused into a patient fresh or frozen. The arguments are stacked in favor of the frozen cellular product, which allows for complete characterization of the therapeutic, including potency evaluation. Applying this procedure, the patient will receive a safe and potent cellular product. Additionally, after performing several experiments in an in vivo engraftment assay we believe that a cryopreserved product also allows for better engraftment. Bone marrow transplant procedures are routinely performed with frozen products and graft failures are rare. Thawing procedures are worked out well in bone marrow transplant units. The thawing process for the experimental gene therapy cellular product follows the same procedure.

Chapter 8
Clinical Applications of HIV Gene Therapy

Before any clinical application of a novel therapy can even be contemplated, many hurdles need to be overcome.

First, it has to be demonstrated in preclinical experiments that the proposed therapy is safe. In the early days of clinical gene therapy, these experiments could often be limited to in vitro experiments. Today, this is not accepted anymore by regulatory agencies, as the past incidences in gene therapy clinical trials warranted much higher scrutiny in preclinical experiments. Safety and toxicity of the novel "cellular drug," which a gene therapy product for the treatment of HIV will in all likelihood be, have to be tested in in vitro experiments in the same cellular product as it will be administered to the patient. Assays have to be available or need to be developed to demonstrate that the transduced cellular product has the same biological properties as the non-transduced cells. Additionally, the product will need to be tested in the appropriate relevant animal model to assure that it will not produce in vivo toxicities. In the United States, such in vitro and in vivo studies will need to be performed under Good Laboratory Practice in accordance to 21CFR Part 58 [109]. Nowadays, great emphasis is placed on the evaluation of genotoxicity and the absence of insertional mutagenesis of the gene therapy product. Experiments therefore will have to be performed to demonstrate this, in vitro and in vivo. The gene therapy vector used for transduction, in the past being a retroviral vector, now a lentiviral vector, needs to undergo safety tests as well. Testing will need to be performed for replication competency, proper sequence, and absence of plasmid DNA that was used to manufacture the vector. Both vector and cellular product will also need to be manufactured under GMP conditions, assuring that the products are free of contaminants and can be manufactured in a highly reproducible fashion. Required additional safety tests after GMP manufacturing are tests for sterility, endotoxin, mycoplasma, and adventitious viruses. This may, however, not be the end for safety testing, as the regulatory agencies may suggest additional tests.

The second aspect that needs to be demonstrated is the efficacy of the novel product that will be slated for clinical applications. In the past, in vitro assays were often sufficient to demonstrate the efficacy of the cellular gene therapy product.

G. Bauer and J.S. Anderson, *Gene Therapy for HIV: From Inception to a Possible Cure*, SpringerBriefs in Biochemistry and Molecular Biology, DOI 10.1007/978-1-4939-0434-1_8, © Gerhard Bauer and Joseph S. Anderson 2014

These assays could be HIV challenge assays performed on anti-HIV gene-expressing target cells; if T cell gene therapy was suggested, these would be T cells, ideally, from HIV-infected individuals, transduced with the clinical grade vector, under conditions mimicking the real clinical scenario. If hematopoietic stem cells are proposed as the gene therapy product, these would be transduced and then matured into macrophages in vitro and finally challenged with HIV. Marked inhibition of HIV output after challenge as compared to cells transduced with a mock vector and non-transduced cells should be achieved; otherwise, the therapeutic value of the product will be doubtful. Nowadays, in vitro assays are not good enough; therefore, animal models for HIV will need to be applied. Over the last few years, it has become possible to engraft immunodeficient mice with human hematopoietic stem cells to generate a functional human immune system, including functional human T and B cells. Such engrafted animals can be challenged, with HIV in vivo, and the in vivo function of anti-HIV gene-expressing HIV target cells can be studied. In such in vivo experiments, the selective survival advantage of anti-HIV gene-expressing T cells in the face of an HIV viral load can be demonstrated.

Keeping all these preclinical issues in mind, in the United States, the following regulatory agencies need to be involved to gain approval for gene therapy clinical trials: The NIH RAC, the Institutional Biosafety Committee (IBC) which is the institutional representative of the NIH RAC, the Institutional Review Board (IRB) which watches over research subject protection, and finally the Food and Drug Administration (FDA), which is the final authority and grants approval of a clinical trial. All regulatory bodies, both local and federal, need to approve the trial; otherwise, it cannot go forward. There are strict reporting mandates to the regulatory bodies; particularly adverse event reporting is vital. The application to the FDA for a clinical trial is called "IND," Investigational New Drug application. There are three phases of clinical trials: Phase I, a safety study; Phase II, a safety and efficacy study; and Phase III, an extended safety and efficacy study. After successful completion of all three phases, an application for marketing of the new drug can be filed with the FDA. Often, a post-marketing clinical study is conducted so the safety and efficacy of the new therapy can be evaluated in a large population and over a long period of time. There is currently no gene therapy application in the marketing phase.

We will now discuss the clinical trials of HIV gene therapy that have already been conducted and will also describe an upcoming new clinical study.

Clinical Trials of T Cell Gene Therapy

The first clinical trial of gene therapy for HIV was a T cell clinical trial using a chimeric CD4+ T cell receptor coupled to a zeta signaling chain which was transferred using a retroviral vector. The hypothesis was that such a receptor transduced into expanded and stimulated peripheral blood T cells (both CD4 and CD8) would bind to cells infected with HIV as the HIV envelope would interact with the chimeric receptor, which in turn would set off a signaling cascade inside the cell, activate it, and cause the infected cell to be killed by the T cell. This was not a definitive anti-HIV gene clinical study; it was more of an adaptive immunity approach. Unfortunately, the clinical efficacy of the study was not high. Two similar studies followed, with the same efficacy results [110, 111].

The next study was a T cell gene therapy study using a definitive anti-HIV gene, the anti-RevM10. A gene gun and gold particles were used to transfer the gene initially. The gene transfer efficiency was very low and cell viability suffered. A follow-up approach was the use of a retroviral vector with higher transduction efficiency. Little clinical efficacy was observed, however [6].

Several other T cell gene therapy studies followed, among them a study transferring a tat ribozyme, a transdominant rev combined with a TAR antisense, and a ribozyme against the U5 leader sequence in HIV. Similar outcomes as outlined above were observed [112–114].

A major change in clinical trials of HIV gene therapy occurred when the first clinical trial using a lentiviral vector was announced. It has to be highly credited to the investigators that they did not shy away from going through the very complicated regulatory process and made the first clinical application of lentiviruses in a gene therapy clinical trial possible. This clinical trial utilized a long antisense molecule directed to the HIV envelope message. In vitro experiments had already shown potent HIV inhibition, and it was hoped that this T cell clinical trial could be used to augment therapy for patients that were resistant to the drugs used in ART. The lentiviral vector manufactured for this trial was a first generation, conditionally mobilizing lentiviral vector which had undergone considerable safety testing. Upon transduction into the target cells, the integrated vector could be packaged into HIV particles after HIV infected such cells. HIV particles could be shown to be highly inhibited by the packaged vector, and they could also spread the anti-HIV vector to other target cells. Several patients were identified that fit the inclusion criteria of multidrug resistance and were treated. Clinical efficacy could be seen in some of the patients which was transient, as expected from a T cell application of HIV gene therapy. There were no adverse events in the clinical trial linked to gene therapy [115].

The latest T cell clinical trial of HIV gene therapy utilizes a zinc-finger nuclease to target and disrupt the CCR5 receptor DNA, preventing its transcription and translation, resulting in a lack of the CCR5 receptor on the cell surface. CCR5 acts as the secondary receptor for the macrophage tropic strains for HIV, and therefore HIV cannot bind and enter the target cell. Additionally, this is not a permanent anti-HIV gene integration strategy, this approach uses a transient exposure of the cells to the zinc-finger nuclease, and no insertional mutagenesis is expected to result from this approach. However, bi-allelic modification is required to really achieve complete knockdown of the CCR5 receptor. This is not always possible. Over 20 patients have already been treated, and clinical efficacy in some of them was reported, among them 1 patient that had a complete drop of the viral load to undetectable levels after being taken off ART. Also an increase in CD4 T cell numbers was observed [116].

As an overall conclusion, it should be pointed out that the T cell gene therapy clinical trials for HIV conducted with either retroviral or lentiviral vectors or the zinc-finger nucleases did not have any serious adverse event linked to gene therapy part. Improvements in clinical efficacy could be seen over time, and it is anticipated that T cell gene therapy for HIV will continue to have therapeutic applications, particularly to augment existing antiretroviral therapies.

Stem Cell Gene Therapy Clinical Trials for HIV

The rationale behind stem cell gene therapy for HIV is the fact that all HIV target cells are derived from hematopoietic stem cells. If anti-HIV genes are permanently inserted into hematopoietic stem cells, these genes will be passed on to their progeny, and after progenitor cell differentiation will be present in mature blood cells. It is hypothesized that even if not all hematopoietic stem cells can be transduced, as selective survival advantage will be conferred onto the anti-HIV gene-expressing mature target cells for HIV, as they will not be killed by HIV. These cells should therefore increase in number in the face of a viral load and contribute to a functional immune system, in spite of HIV infection.

In the 1990s, several clinical trials conducted in Los Angeles, conducted through a collaboration between Children's Hospital Los Angeles and City of Hope, focused specifically on transductions of hematopoietic stem cells using retroviral vectors. Gene transduction was carried out by an expert team at Children's Hospital Los Angeles, while the patients were infused and followed up at City of Hope. In 1996, the first stem cell gene therapy clinical trial for HIV using a tat/rev ribozyme was conducted in five HIV-infected adults that had not progressed to AIDS. Peripheral blood stem cells were mobilized, the patients underwent several apheresis sessions, and their CD34+ cells were isolated from the apheresis products. Fresh CD34+ cells were then transduced using retroviral vectors, over a 3-day period in the presence of growth factors (SCF, IL-6, and IL-3) on irradiated autologous marrow stromal cells (MSCs). It was at that time already that MSCs were cultured under GMP conditions and administered into patients in a clinical trial, in conjunction with their transduced hematopoietic stem cells. Both a control vector transferring only neomycin resistance and a therapeutic vector transferring the anti-HIV ribozyme plus neomycin resistance were used. This was done to see if a selective survival advantage of the anti-HIV gene transduced cells in the face of a viral load would be observable. Gene transduction efficiency into hematopoietic progenitor cells, as measured in colony-forming unit assays in vitro, was about 30–40 %. However, true hematopoietic stem cell transduction efficiency, which cannot be measured in colony assays, must have been low. Gene marking in the peripheral blood, which appeared after about 4 weeks, was also very low, which was expected, as the number of transduced CD34+ cells infused into the patients was small compared to the existing number of CD34+ cells in the bone marrow. There was also no difference in numbers of control gene marked cells versus anti-HIV gene marked cells. There were no adverse events associated with this clinical trial that were related to gene therapy [117].

A groundbreaking clinical trial of stem cell gene therapy for HIV was conducted soon afterward by the same group. Combination HIV drug therapy had just become state of the art at that time; however, HIV lymphoma patients were still not considered eligible for high-dose chemotherapy and autologous bone marrow transplantation since their life expectancy was considered short and insurance companies did not want to take on the cost of such an expensive procedure in this patient population. The researchers believed, however, that HIV lymphoma patients would have a very similar life expectancy as non-HIV-infected individuals with lymphoma if they could be cured of their lymphoma, as they were taking ART. As these patients were in need of a full marrow ablation through high-dose chemotherapy and would need a subsequent autologous bone marrow transplantation to reconstitute their bone marrow, this was thought to be the best population for stem cell gene therapy for HIV as well. The new clinical trial provided a chance to investigate how well anti-HIV gene transduced CD34+ cells would engraft after complete marrow ablation and would perhaps provide a clinical benefit, if enough engraftment could be achieved. Five HIV lymphoma patients were enrolled to be treated for their lymphoma with high-dose chemotherapy and autologous bone marrow transplantation. Peripheral blood stem cells were mobilized after complete remission was achieved by chemotherapy, the patients underwent several apheresis sessions, and then their CD34+ cells were isolated from the apheresis products. Isolated CD34+ cells were then cryopreserved. Half of the apheresis products, however, remained unmanipulated and were also cryopreserved. This was done for safety reasons as no engraftment data on cultured, transduced cells in a fully ablated individual were available. The FDA therefore required that the patients will receive the stored, unmanipulated apheresis product, which could completely reconstitute them, and then, in addition, the transduced CD34+ cells. The patients then underwent high-dose chemotherapy. A few days before the end of the chemotherapy procedure, the cryopreserved CD34+ were thawed, cultured, and transduced, again on autologous MSC in the presence of cytokines. The same anti-HIV gene was used, the tat/rev ribozyme and a control gene, transferred by the same retroviral vectors as before. After chemotherapy, the patients received an infusion of their thawed unmanipulated apheresis product and, a few hours later, an infusion of transduced CD34+ cells. All patients engrafted. Four out of five patients have long-term survival and are free of lymphoma, more than 10 years after the procedure. One patient died of relapse from very aggressive lymphoma. There were no adverse events related to gene therapy, even in a fully marrow-ablated setting. There were no insertional mutagenesis and no genotoxicity. Transduction efficiency, even after thawing of CD34+ cells, was 30–40 %, although a smaller number of CD34+ cells were finally harvested due to initial cell death after thawing. More than enough cells were available for infusion to meet the cell number release criteria. However, the same issue arose as described before, oncoretroviral vectors cannot easily integrate into true stem cells. Initially, there was high gene marking in the peripheral blood, which over a few months diminished and almost completely disappeared by 12 months. This was most likely due to the fact that progenitors initially produced the peripheral blood cells and very few true hematopoietic stem cells were transduced. There was no selective survival advantage for gene transduced peripheral blood T cells, as patients had completely suppressed viral loads since they were required to continue ART, as per FDA request. In spite of the loss of gene marking over the long run, this clinical trial paved the way for future clinical applications. It could be demonstrated that HIV lymphoma patients had practically the same survival rate as non-HIV-infected lymphoma patients; a clinical trial of autologous bone marrow transplantation in HIV lymphoma patients without gene therapy followed and made this procedure the standard of care. It could also be demonstrated that stem cell gene therapy for HIV in a full ablation setting is safe [118].

In 1997, the group at Children's Hospital Los Angeles then pioneered stem cell gene therapy clinical trials in a pediatric patient population. Four children with HIV infection, ages 8–17, were enrolled. All of these patients had viral loads, in spite of ART. Bone marrow aspirates were obtained from these patients, density gradient centrifugation was performed on these aspirates, and CD34+ cells were isolated using immuno-magnetic beads. The isolated cells were then transduced on autologous MSCs for the first three subjects and on RetroNectin for the last subject, with retroviral vectors transferring the neomycin resistance gene, as a control gene and an RRE decoy anti-HIV gene, in the presence of the cytokines SCF, IL-6, and IL-3. Transduction efficiencies as measured by colony-forming unit assays were between 7 and 30 %. The transduced CD34+ cells were infused into the patients without marrow ablation, and as expected, very low gene marking in the peripheral blood resulted. In spite of a viral load in the patients, no selective survival advantage of the anti-HIV gene transduced cells could be observed. There were no adverse events related to the gene transfer. It could again be demonstrated that stem cell gene therapy for HIV is safe and that enough bone marrow CD34+ cells could be isolated and transduced, even for a pediatric population [37].

A reevaluation of the clinical application of anti-HIV genes followed after these initial clinical trials. It was decided at Children's Hospital Los Angeles to perform another pediatric study of stem cell gene therapy for HIV, this time using the strongest anti-HIV gene available at the time, the RevM10 gene, and to select a pediatric population that had a relatively high viral load, in spite of ART. It was thought that a selective survival advantage of gene transduced cells could perhaps be seen, in spite of the fact that the FDA would not allow marrow ablation or the interruption of ART, due to ethical reasons. Two HIV-infected children were enrolled. Bone marrow aspirates were performed, and CD34+ cells were isolated as in the previous study. The transduction conditions were slightly modified, RetroNectin was used as a matrix, and the cytokines SCF, FLt-3, and TPO were used for cell stimulation. The cells were transduced with a RevM10 gene and a non-translated marker gene (FX) as a control. Transduction efficiency was 15–30 %. In both subjects, approximately 1 in 10,000 peripheral blood mononuclear cells were marked, equally with control and anti-HIV gene at around 1–3 months post infusion. After 3 months, vector marking started to decline and disappeared, consistent with what was seen in the previous clinical trials. Both patients continued on ART, their viral load was below 10,000 genomic copies per ml. However, patient 1 had poor adherence to ART, and ART discontinued completely at around 11 months post infusion. The viral load in that patient started to climb, up to 261,000 genomic copies per ml.

Interestingly, at that time, peripheral blood mononuclear cells marked with RevM10 that had previously been undetectable appeared again and were detectable up to 1 in 10,000. The FX gene marked cells, however, did not reappear. For safety reasons, the patient was soon put on a different ART regimen, and the viral load became undetectable. At the same time, the gene marked cells also declined in tight correlation with the viral load and disappeared. The second patient, who had continued on ART, had no reappearance of gene marked cells in the peripheral blood in the same time period. In spite of this being observed in just one patient, selective survival of anti-HIV gene transduced cells in the face of an HIV viral load could be demonstrated in a human clinical trial [39].

There was another very interesting study performed at UCLA, to date the largest randomized clinical trial of stem cell gene therapy for HIV, which enrolled 74 patients. CD34+ cells were transduced with an anti-tat ribozyme or a placebo as a control. CD34+ cells were isolated after mobilization, no conditioning was applied for reinfusion. It was found that the viral load did not significantly decrease in the treatment arm as compared to the placebo arm, but a significant difference was found in the numbers of CD4+ cells. These were significantly higher in the treatment arm. The study therefore demonstrated that anti-HIV gene transduced hematopoietic progenitor cells can produce biologically active CD4+ cells in HIV-infected individuals that have a survival advantage, as the CD4+ cell number correlated with the viral load. Additionally, this large clinical trial demonstrated the safety of this treatment, as no gene therapy-related adverse events were noted [78].

For all the previous clinical trials, oncoretroviral vectors based on the Moloney murine leukemia virus were used. The gene therapy community, at meetings, several times discussed that it would be advantageous to switch to lentiviral vectors for such trials, since better transduction into quiescent hematopoietic stem cells could be achieved. These vectors also offer an improved safety profile as their integration pattern does not favor transcriptional start site of active genes. Another advantage is the fact that they do not shut down gene expression over the long run. However, it took considerable time for all the regulatory steps to be accomplished to start with the first hematopoietic stem cell gene therapy clinical trial for HIV using lentiviral vectors. At City of Hope, a new combination gene therapy vector had been developed. It encompassed a ribozyme to knock down the CCR5 receptor, an shRNA to cut down the tat/rev messenger RNA for HIV, and a TAR decoy. A combination anti-HIV gene strategy is based on the same principle as the combination drug regimen; it prevents the generation of resistant viral mutants. These three anti-HIV genes were packaged into one third-generation lentiviral vector, which was extensively tested preclinically. This vector was different from the first clinical grade lentiviral vector that was used in the T cell gene therapy study that it was a self-inactivating (SIN) vector and could not be mobilized by an HIV particle. The clinical trial enrolled four HIV lymphoma patients, as it was based on the clinical trial model that had been pioneered by the same group previously. CD34+ cells were isolated from apheresis products, transduced with lentiviral vector, and reinfused into a fully marrow-ablated patient, however, again in conjunction with a non-manipulated BACK UP APHERESIS PRODUCT. Transduction efficiencies were lower than those achieved with retroviral vectors previously (1 % long term), and gene marking in the patients was also low (0.02 and 0.32 %, corresponding to 200–3,200 copies per 10e6 PBMCs). However, long-term persistence of gene marking (currently up to 4 years) could be demonstrated, as gene marking did not decline or disappear. This points into the direction of stable transduction of hematopoietic stem cells. Additionally, all subjects have achieved long-term remission of their lymphoma, and no gene therapy-related adverse events were observed [60].

What can be learned from these clinical gene therapy trials? The correct vector choice is very important. Lentiviral vectors are the best vectors for stem cell gene therapy, as they are able to transduce resting cells, do not get shut down, and allow for long-term expression of the anti-HIV genes. Combination anti-HIV genes have to be applied to circumvent the generation of HIV escape mutants. Best would be anti-HIV genes that act prior to reverse transcription to prevent HIV mutation and integration. High transduction efficiency, without increase in copy number per cell (for safety reasons), is a must. Gene marking in the peripheral blood is directly correlated to transduction efficiency. Marrow conditioning is necessary, or low peripheral blood cell marking will be seen, as transduced stem cells will be diluted out in the large number of non-transduced bone marrow stem cells. Finally, after sufficient engraftment of transduced CD34+ cells has been reached, a selective survival advantage must be given to the anti-HIV gene-expressing cells in the peripheral blood to form sufficient numbers of HIV-resistant immune cells. Taken together, all these items may lead to an immune system that could theoretically control HIV infection.

Based on these assumptions, a new clinical trial has been designed at UC Davis Medical Center. A new triple combination anti-HIV gene third-generation lentiviral vector was designed, encompassing an shRNA directed against the CCR5 receptor, a TRIM5 alpha molecule preventing HIV uncoating, and a TAR decoy, preventing upregulation of HIV transcription. A high-titer, clinical grade lentiviral vector was manufactured, and preclinical transduction experiments resulted in transduction efficiencies into CD34+ cells up to 50 %. An HIV in vivo model was developed, and engraftment and in vivo HIV challenge experiments were performed, demonstrating excellent selective survival advantage of anti-HIV gene-expressing peripheral blood cells. A clinical trial was then again designed around the HIV lymphoma population for safety reasons. High-dose chemotherapy will be given to cure the lymphoma; the bone marrow will be reconstituted with only the gene transduced CD34+ cell product. Finally, ART will be withdrawn after stable cell engraftment has been achieved, and enough peripheral blood T cells are seen. This clinical trial has gone through its first regulatory step and has received approval from the NIH RAC [119]. Currently, a pre-IND meeting with the FDA is being prepared. Upon recommendation from the RAC, it should be considered to also expand this new clinical trial to HIV-infected patients without lymphoma, after the protocol has been deemed safe in lymphoma patients.

Chapter 9
Is a Cure for HIV Possible?

In 2007, a bone marrow transplant was performed in Berlin, Germany, on an HIV-infected patient in order to cure his acute myeloid leukemia. This patient, now referred to as the "Berlin patient," received a transplant of hematopoietic stem cells from a donor who was homozygous for a CCR5 delta-32 bp mutation which renders cells deficient of CCR5 surface expression. CCR5, as discussed in previous chapters, is a coreceptor utilized by HIV to attach to and fuse with HIV-susceptible cells during the first stage of the virus' life cycle. This CCR5 delta-32 bp allele is a naturally occurring HIV-resistant genotype which is found in about 1 % of northern European descendants. After the transplant was performed, the patient engrafted and remained free of leukemia. In 2009, an article in the New England Journal of Medicine reported that he had also remained without HIV rebound after being taken off ART and had undetectable viral load for a 20-month follow-up period [120]. Five years after the transplant, the patient still remains without viral load. Many tissue biopsies were taken, and no virus was found. To this date, the "Berlin patient" is the only documented case of an individual being cured of HIV [81].

Experts may question: Can we really cure someone after established HIV infection?

There are two types of cures which should be addressed. The first is called a "sterilizing cure" which is one that completely eradicates HIV from the body of a treated patient. It may be possible that a patient would be transplanted with enough HIV-resistant HSCs that the vast majority of the new immune cells block further HIV infection. Over time, the HIV cellular reservoirs could not be replenished and would die off, as there would not be enough HIV-susceptible cells to reinfect. Thus, the HIV infection would smolder and would cease to exist in a patient.

A sterilizing cure may be extremely difficult to achieve with an integrating virus, however. For instance, the human genome harbors many silenced retroviruses, in the evolutionary path that could never be eradicated, but they were rendered harmless over time. Should we therefore rather define a cure for HIV as a functional cure, with control of the virus by the immune system? Examples for this would be the successful control of EBV or CMV in an immunocompetent individual.

G. Bauer and J.S. Anderson, *Gene Therapy for HIV: From Inception to a Possible Cure*, SpringerBriefs in Biochemistry and Molecular Biology, DOI 10.1007/978-1-4939-0434-1_9, © Gerhard Bauer and Joseph S. Anderson 2014

These viruses persist for the rest of the life of the individual, but they do not cause disease symptoms. Only if the person becomes immunocompromised can these viruses become active and cause disease. An intact immune system eliminates newly replicating viruses and newly infected cells all the time. Could this be achieved in an HIV-infected individual?

A functional cure would most likely be possible when there is a balance between a partially HIV-resistant immune system and a concurrent low-level HIV infection. Patients would be able to stop taking antiretroviral drugs because enough cells in their immune system are HIV-resistant and are capable of maintaining normal immune system functions. Even though there is still a low level of HIV replication, disease progression toward AIDS would not occur due to a subpopulation of HIV-resistant immune cells. The reason why HIV-infected patients progress in their disease to AIDS is because of low CD4+ cell levels which render the body susceptible to opportunistic infections normally controlled by a working immune system. Therefore, if normal CD4+ cell levels are maintained, opportunistic infections can be controlled. It is known that long-term non-progressors and elite controllers who are infected with HIV can have detectable viremia. However, due to their genetic makeup or the strain of virus they were infected with, these individuals are able to control HIV infection. They are able to maintain normal CD4+ cell levels which allow them to not progress to AIDS even though viral replication is still occurring in their body. Also, in certain nonhuman primates including sooty mangabeys who are naturally infected with SIV, the simian form of HIV, they are able to live with ongoing virus replication without developing AIDS-like syndromes. This is likely due to the sooty mangabey's genetics and evolution with the virus which allows them to maintain normal CD4+ cell levels even while SIV is infecting cells and replicating. Therefore, CD4+ cell levels may be a highly critical factor in evaluating an anti-HIV gene during HIV infection. Plasma viremia may not be a definitive result of whether a therapeutic candidate will be successful or not. If maintaining CD4+ cell levels is critical during HIV infection, then a selective survival advantage of anti-HIV gene-modified cells would also be a critical component to evaluate. As HIV-susceptible cells are being infected and killed, the HIV-resistant cells, due to their expression of anti-HIV genes, will survive, proliferate, and take over the immune system. This population of HIV-resistant immune cells would be the contributing factor to maintaining normal CD4+ cell levels above 200 cells/μl and stopping the progression toward AIDS. Therefore, HIV stem cell gene therapy holds considerable promise to deliver a cure for HIV-infected patients.

It has been demonstrated that the CCR5 receptor is vital for HIV infection. Although it is the receptor for just one strain of HIV, the macrophage tropic strain, this strain is the predominant one. Newly acquired HIV infections mostly occur through the mucosa, and only macrophage tropic strains are able to establish these infections. T cell tropic strains arise much later in the natural history of HIV infection, after the immune system is almost destroyed. Therefore, the predominant strain during HIV infection, for a prolonged period of time, is also macrophage tropic. The complete absence of the CCR5 receptor in homozygous individuals is an extremely potent protection from HIV. Even heterozygous individuals for the CCR5 deletion, who still express 50 % of the receptors on the cell surface, are protected from HIV infection, to a degree. The absence of the CCR5 receptor acts as an HIV entry inhibitor, prior to reverse transcription. This fact is another very important aspect, as HIV cannot produce viral mutants. The first line of defense, entry inhibition, is therefore the most important one.

In the "Berlin patient," it has been demonstrated how vital the CCR5 receptor is in controlling HIV. But is there another aspect to the absence of a viral load in that patient?

He received intense chemotherapy that eliminated his hematopoietic system. At the same time, it could have been possible that also the cells that form the HIV reservoir were eliminated, to a large degree. Many HIV patients have undergone high-dose chemotherapy to treat lymphoma, and their viral load did rebound, after successful transplantation. However, there may be a time shortly after chemotherapy where the reservoirs are not plentiful. If HIV cannot enter new cells, as they are resistant to HIV by means of entry inhibition, HIV may not be able to replenish the reservoirs.

Another extremely important lesson to be learned from the "Berlin patient" is the fact that all of the new immune cells he developed lacked the CCR5 receptor. In stem cell gene therapy experiments where only very small numbers of transduced cells were transplanted which then produced an even smaller number of peripheral blood cells that were resistant to HIV, the anti-HIV effect may not have become visible. There may be a threshold of protected cells that needs to be reached before a clinically measurable effect becomes observable.

Is the immune system capable of controlling HIV? This is very likely, if the immune cells cannot be attacked and killed by HIV. We believe this is a numbers game. Only if enough immune cells are available, this will truly be the case. Currently, we do not know an exact number. We only know that if all immune cells are protected from HIV, control of HIV is possible.

Is the "Berlin patient" truly free of HIV? We also do not know the answer to this question. Integrated HIV can hide in very few long-lived cells in hidden sanctuaries. Although many biopsies have been taken and showed no virus, this does not mean that the virus is really eradicated. The current status rather indicates that outstanding control of the virus has been achieved, which in itself is an extremely remarkable accomplishment. Long-term follow-up will show if HIV really has been eradicated.

Can the virus rebound? If the virus, for whatever reason, manages to survive in a sanctuary site and is able to replicate, there is a theoretical possibility that resistant mutants can arise. A huge selective pressure to mutate around the absence of the CCR5 receptor will direct this mutation toward the use of other chemokine receptors, most likely the CXCR4 receptor. Currently, there is no evidence for this in the patient.

How can the lessons learned from the "Berlin patient" be applied to stem cell gene therapy and to make a cure for HIV possible?

If all hematopoietic stem cells can be made resistant to HIV, this would directly follow in the footsteps of the transplantation with bone marrow from a donor with the CCR5 mutation. Therefore, similar results would be expected. This strategy could be pursued if transduced CD34+ could be selected or enriched in vivo. There have been attempts to do this by the insertion of a chemotherapy resistance gene into the CD34+ cells. The results of clinical trials to test this in vivo enrichment method in cancer patients were unfortunately not as good as expected.

Hematopoietic stem cells could be developed from autologous induced pluripotent stem cells (iPSCs) that could be engineered to be resistant to HIV. Integration-free methods to generate iPSCs could be used. The insertion of anti-HIV genes could be controlled and directed into safe harbors in the human genome. The gene-modified iPSCs could be highly tested and selected and then differentiated into hematopoietic stem cells with high purity. This would assure that a highly tested and efficacious product would be transplanted into a patient. Currently, this technology is still under development. The in vivo functionality of iPSC-derived hematopoietic stem cells is very low. After this problem has been solved and efficient production methods have been worked out, this method may become very promising.

Genetically modified autologous human hematopoietic stem cells will have the possibility to cure HIV infection, if highly pure, gene-modified stem cells are transplanted. They will be superior to the transplantation of allogeneic bone marrow stem cells from donors with a CCR5 deletion, as allogeneic transplantation is associated with high morbidity and mortality. To make space in the bone marrow, reduced intensity chemotherapy regimens can be developed for non-cancer patients that will also allow the treatment of older patients. Combination anti-HIV genes can be used for transduction, minimizing the chance for resistant mutants to arise. If cell culture and transduction methods are simplified, this treatment will also become affordable.

In summary, it is very likely that a cure for HIV is possible. If enough funding is made available for quality research in this area and enough translational laboratories make it their priority to apply this research, a cure may become reality for many HIV-infected individuals sooner rather than later.

References

1. Kohn DB (2010) Update on gene therapy for immunodeficiencies. Clin Immunol 135(2): 247–254
2. Fuchs M (2006) Gene therapy. An ethical profile of a new medical territory. J Gene Med 8(11):1358–1362
3. Wolff J et al (2005) Non-viral approaches for gene transfer. Acta Myol 24(3):202–208
4. Melman A et al (2007) Plasmid-based gene transfer for treatment of erectile dysfunction and overactive bladder: results of a phase I trial. Isr Med Assoc J 9(3):143–146
5. Chiarella P, Fazio VM, Signori E (2010) Application of electroporation in DNA vaccination protocols. Curr Gene Ther 10(4):281–286
6. Woffendin C et al (1994) Nonviral and viral delivery of a human immunodeficiency virus protective gene into primary human T cells. Proc Natl Acad Sci U S A 91(24):11581–11585
7. Chalberg TW et al (2006) Gene transfer to rabbit retina with electron avalanche transfection. Invest Ophthalmol Vis Sci 47(9):4083–4090
8. Zolochevska O, Figueiredo ML (2012) Advances in sonoporation strategies for cancer. Front Biosci (Schol Ed) 4:988–1006
9. Plank C et al (2011) Magnetofection™ platform: from magnetic nanoparticles to novel nucleic acid therapeutics. Ther Deliv 2(6):717–726
10. Pozzi D et al (2009) Toward the rational design of lipid gene vectors: shape coupling between lipoplex and anionic cellular lipids controls the phase evolution of lipoplexes and the efficiency of DNA release. ACS Appl Mater Interfaces 1(10):2237–2249
11. Caracciolo G, Amenitsch H (2012) Cationic liposome/DNA complexes: from structure to interactions with cellular membranes. Eur Biophys J 41(10):815–829
12. Wakabayashi T et al (2008) A phase I clinical trial of interferon-beta gene therapy for high-grade glioma: novel findings from gene expression profiling and autopsy. J Gene Med 10(4):329–339
13. Zhou J et al (2011) Systemic administration of combinatorial dsiRNAs via nanoparticles efficiently suppresses HIV-1 infection in humanized mice. Mol Ther 19(12):2228–2238
14. Shaw KL, Kohn DB (2011) A tale of two SCIDs. Sci Transl Med 3(97):97ps36
15. Hacein-Bey-Abina S et al (2010) Efficacy of gene therapy for X-linked severe combined immunodeficiency. N Engl J Med 363(4):355–364
16. Gaspar HB et al (2011) Hematopoietic stem cell gene therapy for adenosine deaminase-deficient severe combined immunodeficiency leads to long-term immunological recovery and metabolic correction. Sci Transl Med 3(97):97ra80
17. Jia H (2006) Controversial Chinese gene-therapy drug entering unfamiliar territory. Nat Rev Drug Discov 5(4):269–270
18. Wilson JM (2009) Lessons learned from the gene therapy trial for ornithine transcarbamylase deficiency. Mol Genet Metab 96(4):151–157

G. Bauer and J.S. Anderson, *Gene Therapy for HIV: From Inception to a Possible Cure*, SpringerBriefs in Biochemistry and Molecular Biology, DOI 10.1007/978-1-4939-0434-1, © Gerhard Bauer and Joseph S. Anderson 2014

19. Wilson RF (2010) The death of Jesse Gelsinger: new evidence of the influence of money and prestige in human research. Am J Law Med 36(2–3):295–325

20. Bennett J et al (2012) AAV2 gene therapy readministration in three adults with congenital blindness. Sci Transl Med 4(120):120ra15

21. Goins WF, Cohen JB, Glorioso JC (2012) Gene therapy for the treatment of chronic peripheral nervous system pain. Neurobiol Dis 48(2):255–270

22. Fletcher JC (1990) Evolution of ethical debate about human gene therapy. Hum Gene Ther 1(1):55–68

23. Friedmann T (2001) Stanfield Rogers: insights into virus vectors and failure of an early gene therapy model. Mol Ther 4(4):285–288

24. Jacobs P (1980) Doctor tried gene therapy on two humans. Washington Post, pp A1, A15

25. Protection of human subjects; reports of the President's Commission for the Study of Ethical Problems in Medicine and Biomedical and Behavioral Research—Office of the Assistant Secretary for Health, HHS. Notice of availability of reports. Fed Regist 48(146):34408–34412 (1983)

26. Walters L (1991) Human gene therapy: ethics and public policy. Hum Gene Ther 2(2): 115–122

27. Recombinant DNA research; actions under guidelines; notice—E. Points to consider for protocols for the transfer of recombinant DNA into the genome of human subjects. Fed Regist 55(41):7443–7447 (1990)

28. Rosenberg SA et al (1990) Gene transfer into humans—immunotherapy of patients with advanced melanoma, using tumor-infiltrating lymphocytes modified by retroviral gene transduction. N Engl J Med 323(9):570–578

29. Culver KW et al (1991) Correction of ADA deficiency in human T lymphocytes using retroviral-mediated gene transfer. Transplant Proc 23(1 Pt 1):170–171

30. Rosenberg SA (1992) Gene therapy of cancer. Important Adv Oncol 17–38

31. Kohn DB et al (1998) T lymphocytes with a normal ADA gene accumulate after transplantation of transduced autologous umbilical cord blood CD34+ cells in ADA-deficient SCID neonates. Nat Med 4(7):775–780

32. Malech HL et al (1997) Prolonged production of NADPH oxidase-corrected granulocytes after gene therapy of chronic granulomatous disease. Proc Natl Acad Sci U S A 94(22):12133–12138

33. Dunbar CE et al (1998) Retroviral transfer of the glucocerebrosidase gene into CD34+ cells from patients with Gaucher disease: in vivo detection of transduced cells without myeloablation. Hum Gene Ther 9(17):2629–2640

34. Cornetta K et al (1992) Retroviral-mediated gene transfer of bone marrow cells during autologous bone marrow transplantation for acute leukemia. Hum Gene Ther 3(3):305–318

35. Brenner MK et al (1993) Gene marking to determine whether autologous marrow infusion restores long-term haemopoiesis in cancer patients. Lancet 342(8880):1134–1137

36. Brenner MK et al (1993) Gene-marking to trace origin of relapse after autologous bone-marrow transplantation. Lancet 341(8837):85–86

37. Kohn DB et al (1999) A clinical trial of retroviral-mediated transfer of a rev-responsive element decoy gene into CD34(+) cells from the bone marrow of human immunodeficiency virus-1-infected children. Blood 94(1):368–371

38. Rossi JJ, June CH, Kohn DB (2007) Genetic therapies against HIV. Nat Biotechnol 25(12):1444–1454

39. Podsakoff GM et al (2005) Selective survival of peripheral blood lymphocytes in children with HIV-1 following delivery of an anti-HIV gene to bone marrow CD34(+) cells. Mol Ther 12(1):77–86

40. Lehrman S (1999) Virus treatment questioned after gene therapy death. Nature 401(6753):517–518

41. Nelson D, Weiss R (2000) Penn researchers sued in gene therapy death: teen's parents also name ethicist as defendant. Washington Post, p A3

42. Cavazzana-Calvo M, Thrasher A, Mavilio F (2004) The future of gene therapy. Nature 427(6977):779–781

43. Zaia JA, Federoff HJ (2009) Death in a gene-therapy trial. N Engl J Med 361(18):1811

44. Demberg T, Robert-Guroff M (2012) Controlling the HIV/AIDS epidemic: current status and global challenges. Front Immunol 3:250
45. Roberts JD, Bebenek K, Kunkel TA (1988) The accuracy of reverse transcriptase from HIV-1. Science 242(4882):1171–1173
46. Huff B (2006) Revisiting monotherapy: heresy or revised orthodoxy? BETA 18(2):15–17
47. Ho DD (1996) Viral counts count in HIV infection. Science 272(5265):1124–1125
48. Siliciano RF, Greene WC (2011) HIV latency. Cold Spring Harb Perspect Med 1(1):a007096
49. Pace MJ et al (2011) HIV reservoirs and latency models. Virology 411(2):344–354
50. Baltimore D (1988) Gene therapy. Intracellular immunization. Nature 335(6189):395–396
51. Geskus RB et al (2007) The HIV RNA setpoint theory revisited. Retrovirology 4:65
52. Hartman TL, Buckheit RW Jr (2012) The continuing evolution of HIV-1 therapy: identification and development of novel antiretroviral agents targeting viral and cellular targets. Mol Biol Int 2012:401965
53. MacGregor RR (2001) Clinical protocol. A phase 1 open-label clinical trial of the safety and tolerability of single escalating doses of autologous CD4 T cells transduced with VRX496 in HIV-positive subjects. Hum Gene Ther 12(16):2028–2029
54. Michienzi A et al (2002) A nucleolar TAR decoy inhibitor of HIV-1 replication. Proc Natl Acad Sci U S A 99(22):14047–14052
55. Bauer G et al (1997) Inhibition of human immunodeficiency virus-1 (HIV-1) replication after transduction of granulocyte colony-stimulating factor-mobilized CD34+ cells from HIV-1-infected donors using retroviral vectors containing anti-HIV-1 genes. Blood 89(7):2259–2267
56. Lee NS et al (2002) Expression of small interfering RNAs targeted against HIV-1 rev transcripts in human cells. Nat Biotechnol 20(5):500–505
57. van Lunzen J et al (2007) Transfer of autologous gene-modified T cells in HIV-infected patients with advanced immunodeficiency and drug-resistant virus. Mol Ther 15(5):1024–1033
58. Anderson J, Akkina R (2007) Complete knockdown of CCR5 by lentiviral vector-expressed siRNAs and protection of transgenic macrophages against HIV-1 infection. Gene Ther 14(17):1287–1297
59. Anderson J, Akkina R (2005) TRIM5alpharh expression restricts HIV-1 infection in lentiviral vector-transduced CD34+-cell-derived macrophages. Mol Ther 12(4):687–696
60. DiGiusto DL et al (2010) RNA-based gene therapy for HIV with lentiviral vector-modified CD34(+) cells in patients undergoing transplantation for AIDS-related lymphoma. Sci Transl Med 2(36):36ra43
61. Anderson J et al (2007) Safety and efficacy of a lentiviral vector containing three anti-HIV genes–CCR5 ribozyme, tat-rev siRNA, and TAR decoy–in SCID-hu mouse-derived T cells. Mol Ther 15(6):1182–1188
62. Vigna E, Naldini L (2000) Lentiviral vectors: excellent tools for experimental gene transfer and promising candidates for gene therapy. J Gene Med 2(5):308–316
63. Deichmann A et al (2011) Insertion sites in engrafted cells cluster within a limited repertoire of genomic areas after gammaretroviral vector gene therapy. Mol Ther 19(11):2031–2039
64. Mantovani J et al (2009) Diverse genomic integration of a lentiviral vector developed for the treatment of Wiskott-Aldrich syndrome. J Gene Med 11(8):645–654
65. Naldini L et al (1996) In vivo gene delivery and stable transduction of nondividing cells by a lentiviral vector. Science 272(5259):263–267
66. Zufferey R et al (1997) Multiply attenuated lentiviral vector achieves efficient gene delivery in vivo. Nat Biotechnol 15(9):871–875
67. Koya RC et al (2002) Transduction of acute myeloid leukemia cells with third generation self-inactivating lentiviral vectors expressing CD80 and GM-CSF: effects on proliferation, differentiation, and stimulation of allogeneic and autologous anti-leukemia immune responses. Leukemia 16(9):1645–1654
68. Whitwam T, Peretz M, Poeschla E (2001) Identification of a central DNA flap in feline immunodeficiency virus. J Virol 75(19):9407–9414

69. Higashimoto T et al (2007) The woodchuck hepatitis virus post-transcriptional regulatory element reduces readthrough transcription from retroviral vectors. Gene Ther 14(17):1298–1304
70. Di Matteo M et al (2012) Recent developments in transposon-mediated gene therapy. Expert Opin Biol Ther 12(7):841–858
71. Schroers R et al (2002) Lentiviral transduction of human T-lymphocytes with a RANTES intrakine inhibits human immunodeficiency virus type 1 infection. Gene Ther 9(13):889–897
72. Zhang MY et al (2010) Potent and broad neutralizing activity of a single chain antibody fragment against cell-free and cell-associated HIV-1. MAbs 2(3):266–274
73. Holt N et al (2010) Human hematopoietic stem/progenitor cells modified by zinc-finger nucleases targeted to CCR5 control HIV-1 in vivo. Nat Biotechnol 28(8):839–847
74. Smyth RP, Davenport MP, Mak J (2012) The origin of genetic diversity in HIV-1. Virus Res 169(2):415–429
75. Zhou J, Rossi JJ (2011) Aptamer-targeted RNAi for HIV-1 therapy. Methods Mol Biol 721:355–371
76. Burnett JC, Rossi JJ (2012) RNA-based therapeutics: current progress and future prospects. Chem Biol 19(1):60–71
77. Zhou J, Rossi JJ (2011) Progress in RNAi-based antiviral therapeutics. Methods Mol Biol 721:67–75
78. Mitsuyasu RT et al (2009) Phase 2 gene therapy trial of an anti-HIV ribozyme in autologous CD34+ cells. Nat Med 15(3):285–292
79. Humeau LM et al (2004) Efficient lentiviral vector-mediated control of HIV-1 replication in CD4 lymphocytes from diverse HIV+ infected patients grouped according to CD4 count and viral load. Mol Ther 9(6):902–913
80. Durand CM et al (2012) HIV-1 DNA is detected in bone marrow populations containing CD4+ T cells but is not found in purified CD34+ hematopoietic progenitor cells in most patients on antiretroviral therapy. J Infect Dis 205(6):1014–1018
81. Allers K et al (2011) Evidence for the cure of HIV infection by CCR5Delta32/Delta32 stem cell transplantation. Blood 117(10):2791–2799
82. Gori JL et al (2012) In vivo selection of autologous MGMT gene-modified cells following reduced-intensity conditioning with BCNU and temozolomide in the dog model. Cancer Gene Ther 19(8):523–529
83. Yam P et al (2006) Ex vivo selection and expansion of cells based on expression of a mutated inosine monophosphate dehydrogenase 2 after HIV vector transduction: effects on lymphocytes, monocytes, and CD34+ stem cells. Mol Ther 14(2):236–244
84. Zhang XY et al (2010) Cell-specific targeting of lentiviral vectors mediated by fusion proteins derived from Sindbis virus, vesicular stomatitis virus, or avian sarcoma/leukosis virus. Retrovirology 7:3
85. Anstee DJ (2010) Production of erythroid cells from human embryonic stem cells (hESC) and human induced pluripotent stem cells (hiPS). Transfus Clin Biol 17(3):104–109
86. Nagata S et al (2009) Efficient reprogramming of human and mouse primary extra-embryonic cells to pluripotent stem cells. Genes Cells 14(12):1395–1404
87. Kambal A et al (2011) Generation of HIV-1 resistant and functional macrophages from hematopoietic stem cell-derived induced pluripotent stem cells. Mol Ther 19(3):584–593
88. Woods NB et al (2011) Brief report: efficient generation of hematopoietic precursors and progenitors from human pluripotent stem cell lines. Stem Cells 29(7):1158–1164
89. Drews K et al (2012) Human induced pluripotent stem cells—from mechanisms to clinical applications. J Mol Med (Berl) 90(7):735–745
90. Joseph A, Sango K, Goldstein H (2009) Novel mouse models for understanding HIV-1 pathogenesis. Methods Mol Biol 485:311–327
91. Brainard DM et al (2009) Induction of robust cellular and humoral virus-specific adaptive immune responses in human immunodeficiency virus-infected humanized BLT mice. J Virol 83(14):7305–7321
92. Shultz LD, Ishikawa F, Greiner DL (2007) Humanized mice in translational biomedical research. Nat Rev Immunol 7(2):118–130

93. Walker JE et al (2012) Generation of an HIV-1-resistant immune system with CD34(+) hematopoietic stem cells transduced with a triple-combination anti-HIV lentiviral vector. J Virol 86(10):5719–5729

94. Stoddart CA et al (2007) Validation of the SCID-hu Thy/Liv mouse model with four classes of licensed antiretrovirals. PLoS One 2(7):e655

95. Bauer G et al (2008) In vivo biosafety model to assess the risk of adverse events from retroviral and lentiviral vectors. Mol Ther 16(7):1308–1315

96. Poropatich K, Sullivan DJ Jr (2011) Human immunodeficiency virus type 1 long-term non-progressors: the viral, genetic and immunological basis for disease non-progression. J Gen Virol 92(Pt 2):247–268

97. Meythaler M et al (2011) Early induction of polyfunctional simian immunodeficiency virus (SIV)-specific T lymphocytes and rapid disappearance of SIV from lymph nodes of sooty mangabeys during primary infection. J Immunol 186(9):5151–5161

98. Hartigan-O'Connor DJ et al (2012) SIV replication in the infected rhesus macaque is limited by the size of the preexisting TH17 cell compartment. Sci Transl Med 4(136):136ra69

99. Zhang J et al (2011) Retroviral restriction factors TRIM5alpha: therapeutic strategy to inhibit HIV-1 replication. Curr Med Chem 18(17):2649–2654

100. Barsov EV et al (2011) Transduction of SIV-specific TCR genes into rhesus macaque CD8+ T cells conveys the ability to suppress SIV replication. PLoS One 6(8):e23703

101. Rosenzweig M et al (1997) Intracellular immunization of rhesus CD34+ hematopoietic progenitor cells with a hairpin ribozyme protects T cells and macrophages from simian immunodeficiency virus infection. Blood 90(12):4822–4831

102. Administration USFAD (2012) 21 Code of Federal Regulations Part 210. Current good manufacturing practice in manufacturing, processing, packing, of holding of drugs; general. http://www.accessdata.fda.gov/scripts/cdrh/cfdocs/cfcfr/CFRSearch.cfm?CFRPart=210&showFR=1

103. Administration USFAD (2012) 21Code of Federal Regulations Part 211. Current good manufacturing practice for finished pharmaceuticals. http://www.accessdata.fda.gov/scripts/cdrh/cfdocs/cfCFR/CFRSearch.cfm?fr=211.196

104. Administration USFAD (2012) 21 Code of Federal Regulations Part 1271. Human cells, tissues, and cellular and tissue based products. http://www.accessdata.fda.gov/scripts/cdrh/cfdocs/cfCFR/CFRSearch.cfm?fr=1271.155

105. CDC (2003) Invasive Streptococcus pyogenes after allograft implantation—Colorado, 2003. MMWR Morb Mortal Wkly Rep 52(48):1174–1176

106. USP 1116: Microbiological evaluation of clean rooms and other controlled environments. http://www.pharmacopeia.cn/v29240/usp29nf24s0_c1116.html

107. Despres D et al (2000) CD34+ cell enrichment for autologous peripheral blood stem cell transplantation by use of the CliniMACs device. J Hematother Stem Cell Res 9(4):557–564

108. Dao MA et al (1998) Adhesion to fibronectin maintains regenerative capacity during ex vivo culture and transduction of human hematopoietic stem and progenitor cells. Blood 92(12):4612–4621

109. Administration USFAD (2012) 21 Code of Federal Regulations Part 58. Good laboratory practice for nonclinical laboratory studies. http://www.accessdata.fda.gov/scripts/cdrh/cfdocs/cfcfr/cfrsearch.cfm?cfrpart=58

110. Walker RE et al (2000) Long-term in vivo survival of receptor-modified syngeneic T cells in patients with human immunodeficiency virus infection. Blood 96(2):467–474

111. Mitsuyasu RT et al (2000) Prolonged survival and tissue trafficking following adoptive transfer of CD4zeta gene-modified autologous CD4(+) and CD8(+) T cells in human immunodeficiency virus-infected subjects. Blood 96(3):785–793

112. Wong-Staal F, Poeschla EM, Looney DJ (1998) A controlled, phase 1 clinical trial to evaluate the safety and effects in HIV-1 infected humans of autologous lymphocytes transduced with a ribozyme that cleaves HIV-1 RNA. Hum Gene Ther 9(16):2407–2425

113. Morgan RA et al (2005) Preferential survival of CD4+ T lymphocytes engineered with anti-human immunodeficiency virus (HIV) genes in HIV-infected individuals. Hum Gene Ther 16(9):1065–1074

114. Macpherson JL et al (2005) Long-term survival and concomitant gene expression of ribozyme-transduced CD4+ T-lymphocytes in HIV-infected patients. J Gene Med 7(5): 552–564

115. Levine BL et al (2006) Gene transfer in humans using a conditionally replicating lentiviral vector. Proc Natl Acad Sci U S A 103(46):17372–17377

116. Autologous CCR5-modified CD4 T-cells effective in HAART nonresponders. AIDS Patient Care STDS 25(11): 693 (2011)

117. Zaia J (1997) Transduction of CD34+ autologous peripheral blood progenitor cells from HIV-1 infected persons: a phase I study of comparative marking using a ribozyme gene and a neutral gene. RAC Protocol Number: 9604-153. http://www.gemcris.od.nih.gov/Contents/GC_CLIN_TRIAL_RPT_VIEW.asp?WIN_TYPE=P&CTID=432

118. Krishnan A (1999) High dose chemotherapy and autologous peripheral stem cell transplantation for HIV lymphomas: a phase IIa study of comparative marking using a ribozyme gene and a neutral gene. RAC Protocol Number: 9710-218. http://www.gemcris.od.nih.gov/Contents/GC_CLIN_TRIAL_RPT_VIEW.asp?WIN_TYPE=P&CTID=364

119. Abedi M (2012) Stem cell gene therapy for HIV in AIDS lymphoma patients. RAC Protocol Number: 1202-1153. http://www.gemcris.od.nih.gov/Contents/GC_CLIN_TRIAL_RPT_VIEW.asp?WIN_TYPE=P&CTID=1156

120. Hutter G et al (2009) Long-term control of HIV by CCR5 Delta32/Delta32 stem-cell transplantation. N Engl J Med 360(7):692–698